D1145247

MATHS IN MINUTES

MATHS IN MINUTES

PAUL GLENDINNING

Quercus

CONTENTS

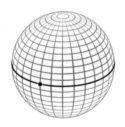 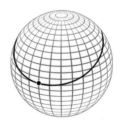

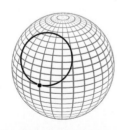

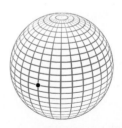

Introduction

Mathematics has been evolving for over four thousand years. We still measure angles using the 360-degree system introduced by the Babylonians. Geometry came of age with the ancient Greeks, who also understood irrational numbers. The Moorish civilization developed algebra and popularized the idea of zero as a number.

Mathematics has a rich history for good reason. It is both stunningly useful – the language of science, technology, architecture and commerce – and profoundly satisfying as an intellectual pursuit. Not only does mathematics have a rich past, but it continues to evolve, both in the sophistication of approaches to established areas and in the discovery or invention of new areas of investigation. Recently computers have provided a new way to explore the unknown, and even if traditional mathematical proofs are the end product, numerical simulations can provide a source of new intuition which speeds up the process of framing conjectures.

Only a lunatic would pretend that all mathematics could be presented in 200 bite-sized chunks. What this book does attempt to do is to describe some of the achievements of mathematics, both ancient and modern, and explain why these are so exciting. In order to develop some of the ideas in more detail it seemed natural to focus on core mathematics. The many applications of these ideas are mentioned only in passing.

The ideas of mathematics build on each other, and the topics in this book are organized so that cognate areas are reasonably close together. But look out for long-range connections. One of the amazing features of mathematics is that apparently separate areas of study turn out to be deeply connected. Monstrous moonshine (page 300) provides a modern example of this, and matrix equations (page 272) a more established link.

This book is thus a heady distillation of four thousand years of human endeavour, but it can only be a beginning. I hope it will provide a springboard for further reading and deeper thought.

Paul Glendinning, Marsden, October 2011.

Numbers

Numbers at their most elementary are just adjectives describing quantity. We might say, for instance, 'three chairs' or 'two sheep'. But even as an adjective, we understand instinctively that the phrase 'two and a half goats' makes no sense. Numbers, then, can have different uses and meanings.

As ancient peoples used them in different ways, numbers acquired symbolic meanings, like the water lily that depicts the number 1000 in Egyptian hieroglyphs. Although aesthetically pleasing, this visual approach does not lend itself to algebraic manipulation. As numbers became more widely used, their symbols became simpler. The Romans used a small range of basic signs to represent a huge range of numbers. However, calculations using large numbers were still complicated.

Our modern system of numerals is inherited from the Arabic civilizations of the first millennium AD. Using 10 as its base (see page 18), it makes complex manipulations far easier to manage.

MMMM   DC   XXII

٤   ٦   ٣٣

4   6   2 2

Natural numbers

Natural numbers are the simple counting numbers (0, 1, 2, 3, 4, …). The skill of counting is intimately linked to the development of complex societies through trade, technology and documentation. Counting requires more than numbers, though. It involves addition, and hence subtraction too.

As soon as counting is introduced, operations on numbers also become part of the lexicon – numbers stop being simple descriptors, and become objects that can transform each other. Once addition is understood, multiplication follows as a way of looking at sums of sums – how many objects are in five groups of six? – while division offers a way of describing the opposite operation to multiplication – if thirty objects are divided into five equal groups, how many objects are in each?

But there are problems. What does it mean to divide 31 into 5 equal groups? What is 1 take away 10? To make sense of these questions we need to go beyond the natural numbers.

One

Together with zero, the number one is at the heart of all arithmetic. One is the adjective for a single object: by repeatedly adding or subtracting the number to or from itself, all the positive and negative whole numbers, the *integers*, can be created. This was the basis of tallying, perhaps the earliest system of counting, whose origins can be traced back to prehistoric times. One also has a special role in multiplication: multiplying any given number by one simply produces the original number. This property is expressed by calling it the *multiplicative identity*.

The number one has unique properties that mean it behaves in unusual ways — it is a factor of all other whole numbers, the first non-zero number and the first odd number. It also provides a useful standard of comparison for measurements, so many calculations in mathematics and science are *normalized* to give answers between zero and one.

Zero

Zero is a complex idea, and for a long time there was considerable philosophical reluctance to recognize and put a name to it. The earliest zero symbols are only found between other numerals, indicating an absence. The ancient Babylonian number system, for instance, used a placeholder for zero when it fell between other numerals, but not at the end of a number. The earliest definitive use of zero as a number like any other comes from Indian mathematicians around the ninth century.

Aside from philosophical concerns, early mathematicians were reluctant to embrace zero because it does not always behave like other numbers. For instance, division by zero is a meaningless operation, and multiplying any number by zero simply gives zero. However, zero plays the same role in addition as one does in multiplication. It is known as the *additive identity*, because any given number plus zero results in the original number.

Infinity

Infinity (represented mathematically as ∞) is simply the concept of endlessness: an infinite object is one that is unbounded. It is hard to do mathematics without encountering infinity in one form or another. Many mathematical arguments and techniques involve either choosing something from an infinite list, or looking at what happens if some process is allowed to *tend to infinity*, continuing towards its infinite limit.

Infinite collections of numbers or other objects, called infinite sets (see page 48), are a key part of mathematics. The mathematical description of such sets leads to the beautiful conclusion that there is more than one sort of infinite set, and hence there are several different types of infinity.

In fact there are infinitely many, bigger and bigger, kinds of infinite set, and whilst this may seem counterintuitive, it follows from the logic of mathematical definitions.

Number systems

A number system is a way of writing down numbers. In our everyday decimal system, we represent numbers in the form 434.15, for example. Digits within the number indicate units, tens, hundreds, tenths, hundredths, thousandths and so on, and are called coefficients. So $434.15 = (4 \times 100) + (3 \times 10) + (4 \times 1) + \left(\frac{1}{10}\right) + \left(\frac{5}{100}\right)$. This is simply a shorthand description of a sum of powers of ten (see page 28), and any real number (see page 22) can be written in this way.

But there is nothing special about this 'base 10' system. The same number can be written in any positive whole-number base n, using coefficients ranging from 0 up to $n - 1$. For example, in base two or binary, the number $8\frac{5}{16}$ can be written as 1000.0101. The coefficients to the left of the decimal point show units, twos, fours and eights – powers of 2. Those to the right show halves, quarters, eighths and sixteenths. Most computers use the binary system, since two coefficients (0 and 1) are easier to work with electronically.

Decimal	Binary
0	0
1	1
2	10
3	11
- - - - - - - - - - - - -	
10	1010
11	1011
12	1100

The number line

The number line is a useful concept for thinking about the meaning of mathematical operations. It is a horizontal line, with major divisions marked by the positive and negative whole numbers stretching away in each direction. The entire range of whole numbers covered by the number line are known as the *integers*.

Addition of a positive number corresponds to moving to the right on the number line by a distance equivalent to the given positive number. Subtraction of a positive number corresponds to moving to the left by that positive distance.

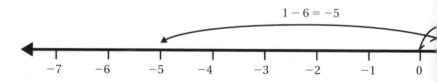

$$1 - 6 = -5$$

Thus one minus ten means moving 10 units to the left of 1, giving minus nine, written -9.

In between the whole number integers shown, there are other numbers, such as halves, thirds and quarters. These are ratios formed by dividing any integer by a non-zero integer. Together with the natural numbers – zero and the positive whole numbers, which are effectively ratios divided by 1 — they form the *rational numbers*. These are marked by finer and finer subdivisions of the number line.

But do the rational numbers complete the number line? It turns out that almost all the numbers between zero and one *cannot* be written as ratios. These are known as *irrational numbers*, numbers whose decimal representations never stop and are not eventually repeating. The complete set of rationals and irrationals together are known as the *real numbers*.

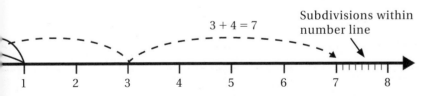

$3 + 4 = 7$

Subdivisions within number line

Families of numbers

Numbers can be classified into families of numbers that share certain properties. There are many ways of putting numbers into classes in this way. In fact, just as there is an infinity of numbers, there is an infinite variety of ways in which they can be subdivided and distinguished from one another. For example the *natural numbers*, whole numbers with which we count objects in the real world, are just such a family, as are the *integers* – whole numbers including those less than zero. The *rational numbers* form another family, and help to define an even larger family, the *irrational numbers*. The families of *algebraic* and *transcendental numbers* (see page 38) are defined by other behaviours while the members of all these different families are members of the *real numbers*, defined in opposition to the *imaginary numbers* (see page 46).

Saying that a number is a member of a certain family is a shorthand way of describing its various properties, and therefore clarifying what sort of mathematical questions we can usefully ask about it. Often, families arise from the creation of functions that describe how to construct a sequence of numbers. Alternatively, we can construct a function or rule to describe families that we recognize intuitively.

For instance we instinctively recognize even numbers, but what are they? Mathematically, we could define them as *all natural numbers of the form* $2 \times n$ *where* n *is itself a natural number*. Similarly, odd numbers are natural numbers of the form $2n + 1$, while prime numbers are numbers greater than 1, whose only divisors are 1 and themselves.

Other families arise naturally in mathematics – for example in the Fibonacci numbers (1, 2, 3, 5, 8, 13, 21, 34, ...), each number is the sum of the previous two. This pattern arises naturally in both biology and mathematics (see page 86). Fibonacci numbers are also closely connected to the golden ratio (see page 37).

Other examples include the times tables, which are formed by multiplying the positive integers by a particular number, and the squares, where each number is the product of a natural number with itself: n times n, or n^2, or n *squared*.

Combining numbers

There are a number of different ways of combining any two given numbers. They can be added together to form their sum, subtracted to form their difference, multiplied together to form their product and divided, provided the divisor is non-zero, to form their ratio. In fact, if we think of $a - b$ as $a + (-b)$ and $\frac{a}{b}$ as $a \times \left(\frac{1}{b}\right)$, then the only operations really involved are addition and multiplication, together with *taking the reciprocal* to calculate $\frac{1}{b}$.

Addition and multiplication are said to be *commutative*, in that the order of the numbers involved does not matter, but for more complicated sequences, the order in which operations are performed can make a difference. To aid clarity in these cases, certain conventions have been developed. Most importantly, operations to be performed first are written in brackets. Multiplication and addition also satisfy some other general rules about how brackets can be reinterpreted, known as *associativity* and *distributivity*, demonstrated opposite.

Commutativity

$$x + y = y + x$$

Associativity

$$(x + y) + z = x + (y + z) = x + y + z$$
$$(xy)z = x(yz) = xyz$$

Distributivity

$$x(y + z) = (xy) + (xz)$$
$$(y + z)x = (yx) + (zx)$$

Rational numbers

Rational numbers are numbers that can be expressed by dividing one integer by another non-zero integer. Thus all rational numbers take the form of fractions or quotients. These are written as one number, the numerator, divided by a second, the denominator.

When expressed in decimal form, rational numbers either come to an end after a finite number of digits, or one or a number of digits are repeated forever. For instance, $0.3333333...$ is a rational number expressed in decimal form. In fraction form, the same number is $\frac{1}{3}$. It is also true to say that any decimal number that comes to an end or repeats must be a rational number, expressible in fractional form.

Since there is an infinite number of integers, it is not surprising to find that there is an infinite number of ways of dividing one by another, but this does *not* mean there is a 'greater infinity' of rational numbers than that of the integers.

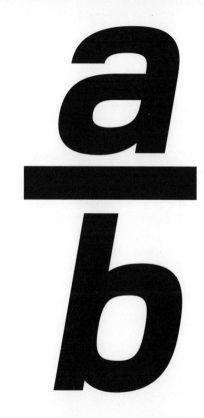

Squares, square roots and powers

The square of any number x is the product of the number times itself, denoted x^2. The term originates from the fact that the area of a square (with equal sides) is the length of a side times itself. The square of any non-zero number is positive, since the product of two negative numbers is positive, and the square of zero is zero. Conversely, any positive number must be the square of two numbers, x and $-x$. These are its *square roots*.

More generally, multiplying a number x by itself n times gives x *to the power of* n, written x^n. Powers have their own combination rules, which arise from their meaning:

$$x^n \times x^m = x^{n+m}, \quad (x^n)^m = x^{nm}, \quad x^0 = 1, \quad x^1 = x, \quad \text{and } x^{-1} = \tfrac{1}{x}.$$

It also follows from the formula $(x^n)^m = x^{nm}$ that the square root of a number can be thought of as that number raised to the power of one-half, i.e. $\sqrt{x} = x^{\frac{1}{2}}$.

Prime numbers

P rime numbers are positive integers that are divisible only by themselves and 1. The first eleven are 2, 3, 5, 7, 11, 13, 17, 19, 23, 29 and 31, but there are infinitely many. By convention, 1 is not considered prime, while 2 is the only even prime. A number that is neither 1 nor a prime is called a *composite number*.

Every composite number can be written uniquely as a product of prime *factors* multiplied together: for example, $12 = 2^2 \times 3$, $21 = 3 \times 7$, and $270 = 2 \times 3^3 \times 5$. Since prime numbers cannot be factorized themselves, they can be thought of as the fundamental building blocks of positive integers. However, determining whether a number is prime, and finding the prime factors if it is not, can be extremely difficult. This process is therefore an ideal basis for encryption systems.

There are many deep patterns to the primes, and one of the great outstanding hypotheses of mathematics, the Riemann hypothesis (see page 396), is concerned with their distribution.

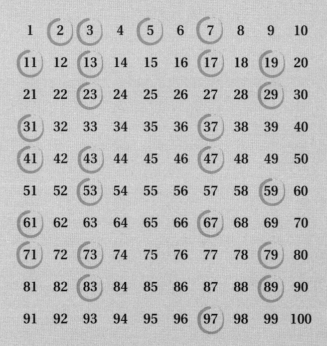

A table of numbers from 1 to 100 with the primes highlighted

Divisors and remainders

A number is a *divisor* of another number if it divides into that number exactly, with no remainder. So 4 is a divisor of 12, because it can be divided into 12 exactly three times. In this kind of operation, the number being divided, 12, is known as the *dividend*.

But what about 13 divided by 4? In this case, 4 is not a divisor of 13, since it divides into 13 three times, but leaves 1 left over. One way of expressing the answer is as *three, remainder one*. This is another way of saying that 12, which is 3×4, is the largest whole number less than the dividend (13) that is divisible by four, and that $13 = 12 + 1$. When the remainder of one is now divided by four, the result is the fraction $\frac{1}{4}$, so the answer to our original question is $3\frac{1}{4}$.

3 and 4 are both *divisors* of 12 (as are 1, 2, 6 and 12). If we divide one natural number, p say, by another, q, that is not a divisor of p, then there is always a remainder, r, that is less

than q. This means that in general $p = kq + r$, where k is a natural number, and r is a natural number less than q.

For any two numbers p and q, the greatest common divisor, *GCD*, also known as the highest common factor, is the largest number that is a divisor of both p and q. Since 1 is obviously a divisor of both numbers, the GCD is always greater than or equal to 1. If the GCD *is* 1, then the numbers are said to be *coprime* – they share no common positive divisors except 1.

Divisors give rise to an interesting family of numbers called 'perfect numbers'. These are numbers whose positive divisors, excluding themselves, sum to the value of the number itself. The first and simplest perfect number is 6, which is equal to the sum of its divisors, 1, 2 and 3. The second perfect number is 28, which is equal to $1 + 2 + 4 + 7 + 14$. You have to wait a lot longer to find the third: 496, which is equal to $1 + 2 + 4 + 8 + 16 + 31 + 62 + 124 + 248$.

Perfect numbers are rare, and finding them is a challenge. Mathematicians have yet to find conclusive answers to some important questions, such as whether there are an infinite number of perfect numbers, or whether they are all even.

Euclid's algorithm

An algorithm is a method, or recipe, for solving a problem by following a set of rules. Euclid's algorithm is an early example, formulated around 300 BC. It is designed to find the greatest common divisor, GCD, of two numbers. Algorithms are fundamental to computer science, and most electronic devices use them to produce useful output.

The simplest version of Euclid's algorithm uses the fact that the GCD of two numbers is the same as the GCD of the smaller number and the difference between them. This allows us to repeatedly remove the larger number in the pair, reducing the size of the numbers involved until one vanishes. The last non-zero number is then the GCD of the original pair.

This method can take many repetitions to reach the answer. A more efficient method, the *standard algorithm*, replaces the larger number by the remainder obtained when dividing it by the smaller number, until there is no remainder.

FINDING THE GCD OF 585 AND 442

Simple version of Euclid's algorithm: 15 steps

$585 - 442 = 143$, so consider 442 and 143
$442 - 143 = 299$, consider 299 and 143
$299 - 143 = 156$, consider 156 and 143
$156 - 143 = 13$, consider 143 and 13
$143 - 13 = 130$, consider 130 and 13

(at this stage the answer is obvious,
but subtracting 13 nine more times leads to ...)

$13 - 13 = 0$, so the GCD is 13.

Standard version of Euclid's algorithm: 3 steps

$$\frac{585}{442} = 1 \text{ (remainder 143)}$$

$$\frac{442}{143} = 3 \text{ (remainder 13)}$$

$$\frac{143}{13} = 11 \text{ (no remainder)}$$

so the process stops, and 13 is the GCD.

Irrational numbers

Irrational numbers are numbers that cannot be expressed by dividing one natural number by another. Unlike rational numbers, they cannot be expressed as a ratio between two integers, or in a decimal form that either comes to an end or lapses into a regular pattern of repeating digits. Instead, the decimal expansions of irrational numbers carry on forever without periodic repetition.

Like the natural numbers and the rationals, the irrationals are infinite in extent. But whilst the rationals and the integers are sets of the same size, or *cardinality* (see page 56), the irrationals are far more numerous still. In fact their nature makes them not only infinite, but uncountable (see page 64).

Some of the most important numbers in mathematics are irrational, including π, the ratio between the circumference of a circle and its radius, Euler's constant e, the golden ratio shown opposite, and $\sqrt{2}$, the square root of 2.

The golden ratio is the ratio between two numbers when the ratio of the smaller one to the larger is equal to the ratio of the larger to the sum of the whole. It is an irrational number and a constant that arises naturally in many situations and is used to govern proportion in art and architecture.

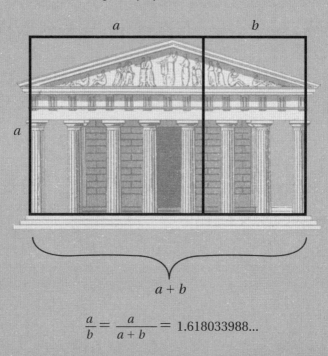

$$\frac{a}{b} = \frac{a}{a+b} = 1.618033988...$$

Algebraic and transcendental numbers

An algebraic number is one that is a solution to an equation involving powers of the variable x, a polynomial (see page 184) with rational coefficients, while a transcendental number is one that is not such a solution. The coefficients in such equations are the numbers that multiply each of the variables. For example, $\sqrt{2}$ is irrational, since it cannot be written as a ratio of two whole numbers. But it *is* algebraic, since it is the solution of $x^2 - 2 = 0$, which has rational coefficients (1 and 2). All rational numbers are algebraic, since any given ratio $\frac{p}{q}$ can be found as the solution of $qx - p = 0$.

We might expect transcendental numbers to be rare, but in fact the opposite is true. $\sqrt{2}$ is exceptional, and almost all irrationals are also transcendental. Proving this is very difficult, but a randomly chosen number between zero and one would almost certainly be transcendental. This raises the question of why mathematicians spend so much time solving algebraic equations, ignoring the vast majority of numbers.

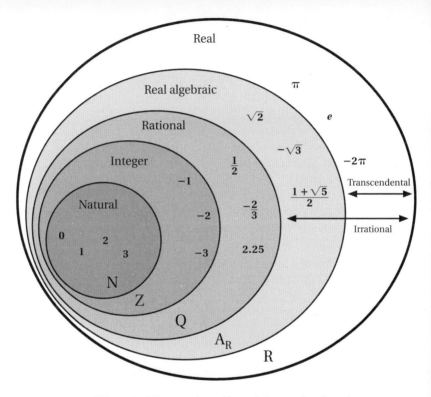

This nested diagram shows the main types of real number, including some important examples.

π

π is a transcendental number and one of the fundamental constants of mathematics. Represented by the Greek letter π, it turns up in a variety of different and unexpected places. It is so important that some mathematicians and computer scientists have devoted a great deal of time and effort towards calculating it ever more precisely. In 2010 the largest number of decimal places reported to have been calculated, using a computer of course, was over 5 trillion!

For all practical purposes, such precision is unnecessary, and π can be approximated by rational numbers $\frac{22}{7}$ and $\frac{355}{113}$, or in decimal notation, by **3.1415926535897932384626433**8. It was first discovered through geometry, perhaps as early as 1900 BC in Egypt and Mesopotamia, and is usually introduced as the ratio of the circumference of a circle to its diameter. Archimedes used geometry to find upper and lower bounds for this value (see page 92), and it has since been found to appear in fields as apparently unrelated as probability and relativity.

3.141592653589793238462643383279502884197 16
9399375105820974944592307816406286208998628
0348253421170679821480865132823066470938446
0955058223172535940812848111745028410270193
8521105559644622948954930381964428810975665
9334461284756482337867831652712019091456485
6692346034861045432664821339360726024914127
3724587006606315588174881520920962829254091
7153643678925903600113305305488204665213841
4695194151160943305727036575959195309218611
7381932611793105118548074462379962749567351
8857527248912279381830119491298336733624406
5664308602139494639522473719070217986094370
2770539217176293176752384674818467669405132
0005681271452635608277857713427577896091736
3717872146844090122495343014654958537105079
2279689258923542019956112129021960864034418
1598136297747713099605187072113499999983729
7804995105973173281609631859502445955...

e

e is a transcendental number and one of the fundamental constants of mathematics. Known as Euler's constant, it has a value of approximately **2.71828182845904523536028747**. Its natural home is in mathematical analysis, and although engineers and physicists are happy to work with powers of ten and logarithms (see page 44) to the base ten, mathematicians almost always work with powers of *e* and logarithms to the base *e*. These are known as natural logarithms.

Like π, *e* has many definitions. It is the unique real number for which the derivative (see page 208) of the function e^x, the *exponential function*, is itself. It is a natural proportion in probability; and it has many representations in terms of infinite sums.

e is intimately related to π, since trigonometric functions (see page 200), which are often expressed using π, can also be defined using the exponential function.

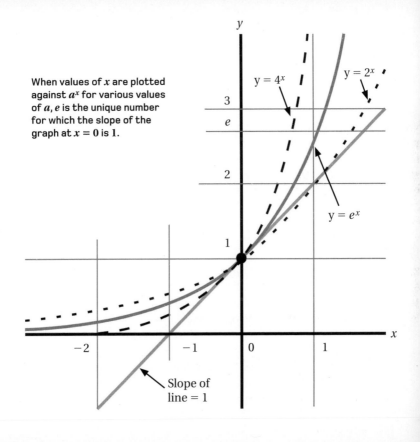

When values of x are plotted against a^x for various values of a, e is the unique number for which the slope of the graph at $x = 0$ is 1.

$y = 4^x$

$y = 2^x$

$y = e^x$

Slope of line $= 1$

Logarithms

L ogarithms are a useful way of measuring the order of magnitude of a number. The logarithm of a number is the power to which a fixed number, the base, must be raised in order to produce the given number. If a given number b can be expressed as 10^a then we say that a is the *logarithm to base* 10 of b, denoted $\log(b)$. Since the product of a number raised to different powers can be obtained by adding those powers, we can also use logarithms to achieve any multiplication involving powers.

Thus by setting $a^n = x$ and $a^m = y$, the rule $a^n a^m = a^{n+m}$ can be written in logarithmic form as $\log(xy) = \log(x) + \log(y)$, while $(a^n)^w = a^{nw}$ is $\log(x^w) = w\log(x)$.

These rules were used to simplify large calculations in an era before electronic calculators, by using logarithm tables or slide rules – two rulers with logarithmic scales that move against each other, where addition of the scales implies multiplication.

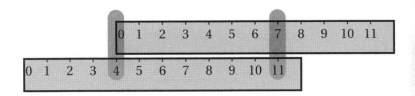

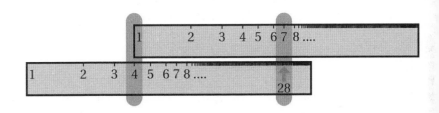

Diagrams of a proportional and a logarithmic slide rule. In a proportional slide rule, aligning the numbers to be added, in this case 4 and 7, as shown reveals the sum. A logarithmic slide is used for multiplication aligning numbers in the same way reveals the product.

i

---

i is a 'number' used to represent the square root of -1. This otherwise unrepresentable concept is not really a number in the sense of counting, and is known as an imaginary number.

The concept of i is useful when we are trying to solve an equation like $x^2 + 1 = 0$, which can be rearranged as $x^2 = -1$. Since squaring any positive or negative real number always gives a positive result, there can be no real-number solutions to this equation. But in a classic example of the beauty and utility of mathematics, if we define a solution and give it a name (i), then we can reach a perfectly consistent extension of the real numbers. Just as positive numbers have both a positive and negative square root, so $-i$ is also a square root of -1, and the equation $x^2 + 1 = 0$ has two solutions.

Armed with this new imaginary number, a new world of complex numbers, with both real and imaginary components, opens out before us (see pages 288–311).

$$\sqrt{-1}$$

Introducing sets

A set is simply a collection of *objects*. The objects within a set are known as its *elements*. The idea of the set is a very powerful one, and in many ways sets are the absolutely fundamental building blocks of mathematics – more basic even than numbers.

A set may have a finite or infinite number of elements, and is usually described by enclosing the elements in curly brackets { }. The order in which the elements are written does not matter in the specification of the set, nor does it matter if an element is repeated. Sets may also be made up from other sets, though great care must be taken in their description.

One reason sets are so useful is because they allow us to retain generality, putting as little structure as possible onto the objects being studied. The elements within a set can be anything from numbers to people to planets, or a mix of all three, although in applications elements are usually related.

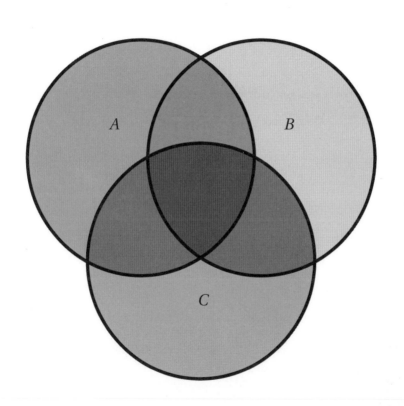

Combining sets

Given any two sets, we can use various *operations* to create new sets, several of which have their own shorthand.

The *intersection* of two sets X and Y, written as $X \cap Y$, is the set of all elements that are members of both X and Y, while the *union* of X and Y, written as $X \cup Y$, is the set of all elements that are in at least one of the sets X and Y.

The *empty set,* represented as $\{\}$ or $\varnothing$, is the set that contains no elements at all. A *subset* of a set X is a set whose elements are all within X. It may include some or all elements of X, and the empty set is also a possible subset of any other set.

The *complement* of Y, also known as *not Y* and written $\overline{Y}$, is the set of elements in not in Y. If Y is a subset of X, then the *relative complement* of Y, written $X \setminus Y$, is the set of elements in X that are not in Y, and this is often referred to as X *not* Y.

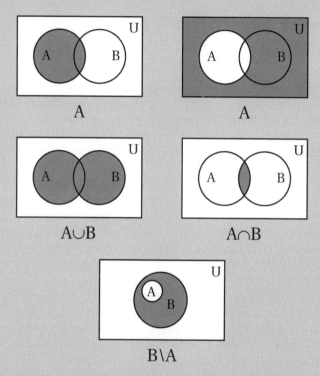

Simple Venn diagrams (see page 52) for some basic set operations

Venn diagrams

Venn diagrams are simple visual diagrams widely used to describe the relationships between sets. In their simplest form, a disc is used to represent each set, and the intersections of discs denote the intersections of sets.

The use of such diagrams for representing the relationships between different philosophical propositions or different sets goes back centuries. It was formalized by British logician and philosopher John Venn in 1880. Venn himself referred to them as *Eulerian circles* in reference to a similar kind of diagram developed by the Swiss mathematician Leonhard Euler in the 18th century.

For three sets, there is a classical way of showing all the possible relationships (see page 49). But for more than three sets, the arrangement of intersections rapidly becomes much more complex. The diagram opposite shows one approach to linking six differerent sets.

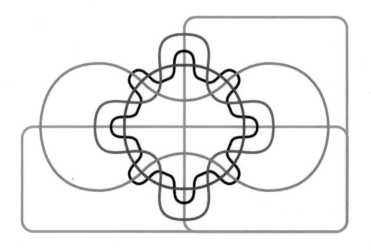

One possible solution for representing
six sets in a Venn diagram.

The barber paradox

A paradox is a seemingly true statement that contradicts itself, or leads to a situation that seems to defy logic. In 1901, British mathematician Bertrand Russell used the barber paradox to expose the flaws in elementary set theory:

All the men in a village either shave themselves or are shaved by a barber (himself a man from the village). The barber claims to shave only the male villagers who do not shave themselves. So who shaves the barber?

Restated in terms of sets, the paradox asks us to consider a set containing all those subsets which do not have themselves as an element. Is this set an element of itself? The immediate solution to such paradoxes was to restrict set theory with a series of rules or *axioms*, creating a hierarchy of sets that are allowed to be elements only of sets above them in the hierarchy. Although not the most elegant of solutions, axiomatic set theories have become widely accepted.

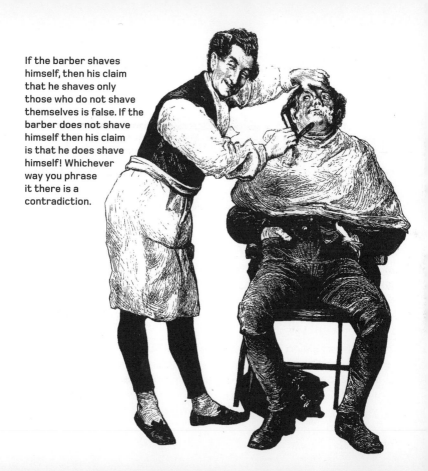

If the barber shaves himself, then his claim that he shaves only those who do not shave themselves is false. If the barber does not shave himself then his claim is that he does shave himself! Whichever way you phrase it there is a contradiction.

Cardinality and countability

The cardinality of a finite set A, written $|A|$, is the number of distinct elements within it. Two sets, whether finite or infinite, are said to have the same cardinality if their elements can be put into one-to-one correspondence. This means that elements of each set can be paired off, with each element associated with precisely one element in the other set.

Countable sets are those sets whose elements can be labelled by the natural numbers. Intuitively, this means that the set's elements can be listed, although the list may be infinite. Mathematically, it means the set can be put into one-to-one correspondence with a subset of the natural numbers.

This has surprising consequences. For instance, a strict subset of a countable set can have the same cardinality as the set itself. So, the set of all even numbers has the same cardinality as the set of square numbers, which has the same cardinality as the natural numbers. All are said to be *countably infinite*.

Hilbert's hotel

Hilbert's hotel is an analogy invented by mathematician David Hilbert in order to visualize the strange idea of countable infinities. This imaginary hotel has a countably infinite set of rooms numbered $1, 2, 3, ...$, and is fully occupied, when a latecomer arrives and pleads for a room.

After some thought, the concierge uses a tannoy system to ask every guest to move into the next room up in numerical order. So the occupant of room 1 moves into room 2, room 2 moves to room 3 and so on. For any of the, countably infinite, guests in room N, there is always a room $N + 1$ for them to move into, so that by the time everyone has moved, room 1 is free for the new guest to occupy.

Hilbert's hotel shows that the result of adding an element to a countably infinite set is still a countably infinite set, so there must be different countable infinities.

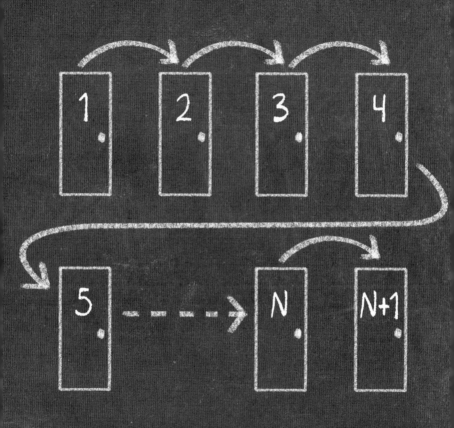

Counting rational numbers

Although not all infinite sets are countable, some very big sets are. These include the rational numbers – numbers made from a ratio of two integers $\frac{a}{b}$. We can prove this by looking at just the rationals between 0 and 1.

If the rationals between 0 and 1 are countable, then we should be able to put them in an order that creates a complete, if infinite, list. The natural ascending order of size is unhelpful here because between any two rational numbers one can always find another, so we could not write down even the first and second elements of such a list. But is there another way to list the numbers?

One solution is to order the numbers by their denominator, b, first, and then by the numerator a, as shown opposite. There is some repetition in this approach, but each rational number between 0 and 1 will appear at least once in the list.

$1/2$

$1/3$, $2/3$

$1/4$, $2/4$, $3/4$

$1/5$, $2/5$, $3/5$, $4/5$

$1/6$, $2/6$, $3/6$, $4/6$, $5/6$

$- - - - - - - - - - - - - - - -$

$1/n$, $2/n$, $3/n$, $4/n$, ..., $(n-2)/n$, $(n-1)/n$

Dense sets

Density is a property that describes relationships between sets and their subsets when there is a notion of a distance between elements of the sets. It provides a way of assessing the relative 'size' of different infinite sets that is different from counting the elements. For instance, one way to make mathematical sense of the idea that the rational numbers are a 'very big' set, is that they are *dense* within a specific subset, in this case the real numbers, which are 'very big' themselves.

A set X is said to be *dense in another set Y*, if X is a subset of Y, and any point in X is either an element of Y, or arbitrarily close to one: for any point in Y we can choose any distance d greater than 0 and find a point in X within distance d of that point.

To prove that the rationals are *dense in the reals*, for example, we select a distance d and a real number y, then prove that there is always a rational number x within d of y, which can be done by truncating the decimal expansion of y.

$$\frac{1}{4} \quad \frac{1}{3} \quad \frac{2}{5} \quad \frac{5}{11} \quad \frac{1}{2} \quad \frac{6}{11} \quad \frac{3}{5} \quad \frac{2}{3}$$

Uncountable sets

Uncountable sets are infinite sets whose elements cannot be arranged in a countable order. The existence of such sets means that there are at least two types of infinite set, countable and uncountable, and it turns out that there are infinitely many different types of uncountable set.

How can we prove if a set is countable? In 1891, German mathematician Georg Cantor used proof by contradiction to show that the set of real numbers between 0 and 1 is uncountable. If it *is* countable, he reasoned, then there is an infinite but countable list of its elements, each of which can be written in the form:

$$0.a_1 a_2 a_3 a_4 \ldots$$

where each digit a_k is a natural number between 0 and 9.

Cantor contradicted this statement by showing that it is always possible to construct a real number between 0 and 1

that is not in this list. Suppose that the kth real number on the list has the decimal expansion:

$$0.a_{k1}a_{k2}a_{k3}a_{k4}\ldots$$

In that case, we can create a number *not* on the list by looking at the first number in the list, $k = 1$, and choosing the first digit in the decimal expansion of our new number as 7 if $a_{11} = 6$, and 6 otherwise. To choose the second digit, we apply the same rule, but using the second digit of the list's second number. The third digit is found from the third number, and so on:

$$0.a_{11}a_{12}a_{13}a_{14}\ldots$$

$$0.a_{21}a_{22}a_{23}a_{24}\ldots$$

$$0.a_{31}a_{32}a_{33}a_{34}\ldots$$

At the end of this infinite process we would have a number whose decimal expansion involves only the digits 6 and 7 and which differs from any nth entry on the list in the nth decimal place – so the original list is not complete, and the set is uncountable. This is known as Cantor's diagonal argument.

Cantor sets

Cantor sets are the earliest appearance of the objects known as fractals. The diagonal argument developed by Georg Cantor (see page 64) shows that certain intervals of the real number line are uncountable sets. But do all uncountable sets contain such line intervals? Cantor showed that it was possible to construct an uncountable set that contains no line intervals. Cantor sets are *infinitely intricate*; they have structure on finer and finer scales.

One example is called the *middle third Cantor set*. It is obtained by starting with an interval and removing the middle thirds from all the intervals remaining at each stage. At the nth stage of building, it has 2^n intervals, each of length $\frac{1}{(3^n)}$, and a total length of $\left(\frac{2}{3}\right)^n$. As n tends towards infinity, so does the number of points within it, while the length of the set shrinks towards zero. It takes a bit more work to show that there really is something left at the infinite limit of this subdivision and to prove that the set is uncountable, but it can be done.

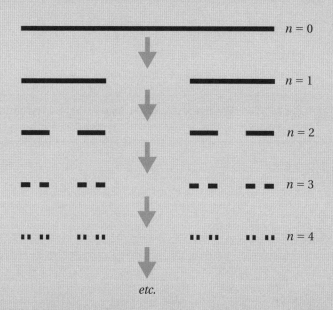

Constructing the Cantor set:
Begin with the *closed unit interval*, the real numbers between 0 and 1 including the end-points, and remove the middle third, leaving two closed intervals of length $\frac{1}{3}$, including their end-points. Now remove the middle third of each of these intervals, so we have four (2^2) closed intervals, each of length $\frac{1}{9}$ ($\frac{1}{3^2}$). Repeat the process *ad infinitum*.

Hilbert's problems

Hilbert's problems are a list of 23 mathematical research problems set out by David Hilbert at the 1900 International Congress of Mathematics in Paris. He considered them to be key to the development of mathematics in the 20th century.

Throughout the 1800s, the axiomatic system, first used by Euclid of Alexandria (see page 108) had been applied in many new areas. Mathematicians had developed methods for finding the defining axioms of the area being studied, for example in geometry, points, lines, curves and their properties, and then developing the subject from these axioms through logic.

Many of Hilbert's problems related to the extension of the axiomatic method, and their solutions advanced mathematics significantly, although the work of Kurt Gödel (see page 70) soon changed the way axiomatic theories themselves were viewed. They also established a fashion for setting lists of mathematical puzzles that continues to this day.

'History teaches the continuity of the development of science. We know that every age has its own problems, which the following age either solves or casts aside as profit-less and replaces by new ones.'

David Hilbert

Gödel's incompleteness theorems

Gödel's incompleteness theorems are two remarkable results that changed how mathematicians view axiomatic mathematics. Developed by German mathematician Kurt Gödel in the late 1920s and early 1930s, the theorems grew out of his method for coding statements in axiomatic theories, and for showing how statements could be modified by logical rules.

Although the axiomatic method for describing various fields of mathematics had proved highly successful, some theories had been shown to require infinite sets of axioms in themselves, and therefore mathematicians were anxious to find formal ways of proving the completeness and consistency of a given set of axioms.

A set of axioms is held to be *complete* if it is capable of proving or negating any statement given in its appropriate language, while a set of axioms is *consistent* if no statement can be made that can be both proved and negated.

Gödel's first theorem states that:

In any (appropriate) axiomatic theory, there exist statements which make sense within the theory but which cannot be proved true or false within that theory.

This means that the axioms of a theory, which we might hope to describe that theory completely, can never do this, and that it is always possible to augment the number of axioms.

As if this wasn't bad enough, a second complication involved the internal consistency of sets of axioms:

It is only possible to prove that an (appropriate) set of axioms is inconsistent, and not that they are consistent.

In other words, we can never be sure that a set of axioms does not contain a hidden contradiction.

Gödel's results have profound implications for the philosophy of mathematics – but, in general, working mathematicians tended to carry on as though nothing had changed.

The axiom of choice

The axiom of choice is a fundamental rule that is often added to the list of axioms used to define mathematical thinking. It is used implicitly in Cantor's diagonal argument (see page 64), and many other mathematical proofs that involve assuming infinite lists have some abstract existence, and that an infinite set of choices can be made.

More precisely, these proofs state that, given an infinite number of non-empty sets containing more than one element, it is possible to choose an infinite sequence of elements with precisely one from each set. To some this seems absurd – infinity rearing its awkward head again – but the rule allowing such a procedure is the axiom of choice.

Other axioms can be chosen, which allow the axiom of choice to arise as a theorem, but whichever version is used, this addition to the basic set of logical rules is necessary to make such arguments permissible.

Probability theory

Probability is the branch of mathematics that deals with measuring and predicting the likelihood of certain outcomes. It is both an application of set theory, and an entirely new theory in itself.

One way of looking at probabilities is by treating a range of possible results as elements of a set. Take for example the case of a fair coin tossed three times. The set of all possible outcomes can be represented with elements consisting of three letters, one per coin-toss, with H standing for heads and T for tails. Clearly this set has eight elements:

$$\{TTT, TTH, THT, THH, HTT, HTH, HHT, HHH\}$$

Since one of these outcomes must occur, the sum of all these probabilities must be 1, and if the coin is fair and each outcome is equally probable, the likelihood of each case is $\frac{1}{8}$.

More complicated questions about probabilities can be answered by considering specific outcomes as sets that are subsets of the previous set of all possible outcomes.

For example, we can see immediately that the set of outcomes with precisely two heads contains three elements, and so has a probability of $\frac{3}{8}$.

But what about the probability that precisely one throw is a head given that at least one is a tail? If we know that at least one throw is a tail we can restrict the set of outcomes to:

$$\{TTT, TTH, THT, THH, HTT, HTH, HHT\}$$

Two elements of this set, out of seven in total, have precisely one head – so the probability is $\frac{2}{7}$.

Similar but more generalized arguments have allowed mathematicians to develop a set of axioms for probability, written in terms of the probability of sets and the operations defined on sets.

Power sets

The power set of a given set S is the set of all subsets of S, including S itself and the empty set. So if $S = \{0, 1\}$, then its power set, denoted $P(S)$ is $\{\emptyset, \{0\}, \{1\}, \{0, 1\}\}$.

The German mathematician Georg Cantor used the power set to show that there are infinitely many different bigger and bigger classes of infinity, using an argument somewhat similar to, though predating, the barber paradox (see page 54).

Cantor's diagonal argument (see page 64) had already shown that there were at least two types of infinite set – countable, or listable, ones, and uncountable sets such as the *continuum*, the set of real numbers. Cantor now showed that if S is an infinite set then its power set will always be bigger than S, in the sense that there is no way to map the elements of S to the elements of $P(S)$ so that each element in one set is associated with one and only one element of the other set. In other words, the cardinality of $P(S)$ is always larger than that of S itself.

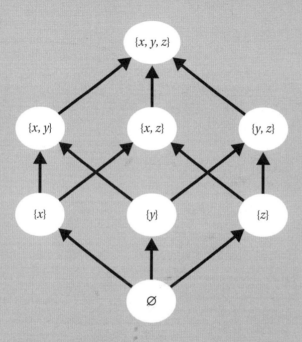

This diagram shows the hierarchy of subsets within the power set of a set {x, y, z} Arrows indicate where subsets of $P\{x, y, z\}$ are also subsets of other subsets.

Introducing sequences

Mathematical sequences are ordered lists of numbers. Like sets (see page 48), sequences can be never-ending, or infinite. Unlike sets, the elements or terms within a sequence have a specific order, and the same terms can recur at different points within the list.

The most familiar sequences are lists of natural numbers, such as $1, 2, 3, \ldots,$. The terms in this sequence are evenly spaced and continue towards infinity. A variant is the Fibonacci sequence, in which the space between terms grows larger. Both are divergent sequences. Other sequences are convergent, closing in on a specific value as they approach the limit of infinite terms.

The terms in a sequence representing radioactive decay, where the remaining quantity of a radioactive isotope halves over a regular interval called the 'half life', get closer to zero as the sequence progresses. This convergent sequence can be illustrated by an exponential decay curve, as shown opposite.

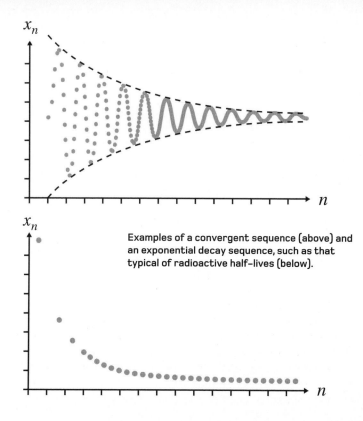

Examples of a convergent sequence (above) and an exponential decay sequence, such as that typical of radioactive half-lives (below).

Introducing series

A mathematical series is an expression for the sum of the terms within a sequence. Usually indicated by the Greek letter Σ (sigma), a series can be the sum of an infinite number of terms, or of a limited range. In each case, the range's lower and upper limits are appended below and above the Σ sign.

Given any sequence of numbers (a_n), the series is the infinite sum:

$$\sum_{i=0}^{\infty} a_i = a_0 + a_1 + a_2 + a_3 + \ldots$$

In many cases, this sum will tend towards infinity or it may not settle down close to a particular value. However, there are series in which the sum tends towards a single number, known as the limit. To see if a series has a meaningful limit, we define the finite partial sum S_n as the sum of the first $n + 1$ terms, $a_0 + a_1 + \ldots + a_n$. The series will converge to a limit L if the associated sequence of partial sums for each n tends towards L.

$$S_1 = 1$$

$$S_2 = 1 + \frac{1}{2} \qquad\qquad 2 - \frac{1}{2}$$

$$S_3 = 1 + \frac{1}{2} + \frac{1}{4} \qquad\qquad 2 - \frac{1}{4}$$

$$S_4 = 1 + \frac{1}{2} + \frac{1}{4} + \frac{1}{8} \qquad\qquad 2 - \frac{1}{8}$$

$$- - - - - - - - - - - - - - - - - -$$

$$S_n = 1 + \frac{1}{2} + \cdots + \frac{1}{2^n} \qquad\qquad 2 - \frac{1}{2^{n-1}}$$

Limits

The limit of an infinite sequence or series, if it exists, is the single value approached as the number of terms in that list or sum tends to infinity. The process of taking limits allows us to make sense of the infinite process by taking a series of approximations and then determining whether the sequence of answers approaches ever closer to a single answer.

Taking limits is an important way of dealing with never-ending or infinite processes, and is absolutely fundamental to mathematics. Though it was used by the Greeks, to calculate approximations of π among other things, and by Isaac Newton, it was not fully formalized until the late 19th century.

Now the backbone of many areas of mathematics, limits are principally used in the field of analysis (see page 208), when studying mathematical functions, the relationships between variables or the development of calculus.

Zeno's paradox

Zeno's paradox is one of several paradoxes posed by Greek mathematician Zeno of Elea, in the fifth century BC:

Tortoise and Hare are running a race over a two-mile track. Hare runs off at a steady pace. Tortoise, being a philosophical creature, sits down, safe in the knowledge that Hare will never arrive at the finishing line.

First, thinks Tortoise, Hare has to run a mile, then half the final mile, then half the final half-mile, and so on. Surely it is impossible for Hare to cover this infinite number of distances?

Zeno's paradox raises both mathematical and philosophical issues. From a mathematical point of view the key point is that, in some cases, infinite sequences of numbers produce summed series that converge to a finite value, so if this is true for the distance covered and the time taken to cover the finite distance, then the hare shoud arrive without any problems.

The Fibonacci sequence

The Fibonacci sequence is a simple pattern created by adding two numbers to make a third. Named after the Italian mathematician who introduced it to the West in 1201, it appears in several areas of mathematics, and also in observations of the physical and natural world.

In mathematical terms the sequence is defined as:

$$F_{n+1} = F_n + F_{n-1} \quad \text{(with } F_0 = 0 \text{ and } F_1 = 1\text{)}.$$

The rule results in a chain of numbers beginning: 0, 1, 1, 2, 3, 5, 8, 13, 21, 34, 55, 89, ...,. In biology, these numbers are mirrored in the relationship between the twists of a plant stem and the number of leaves along the stem, in the spiral arrangement of sunflower seeds, and in many other naturally-occurring patterns. The Fibonacci sequence is also useful in a range of mathematical contexts, including the solution of Euclid's algorithm. It is also linked with the golden ratio (see page 37).

Convergent sequences

The terms in an ordered list of numbers are convergent if they progressively close in on a specific value or limit. But, while we may observe that a sequence seems to be converging on a limit, how can we know what that limit is? For example, methods of estimating π often rely on a sequence approach. As the sequence gets closer and closer to a number, it would be nice to say that this is the true value of π.

If a number L is known, then a sequence tends to L if, given any level of error ε, there is some stage of the sequence after which all the remaining terms are within ε of L. Karl Weierstrass and others discovered that it was not necessary to know L in order to determine whether a sequence converges.

A *Cauchy sequence* is one in which, given any level of error ε, there is some stage in the sequence after which any two points remaining in the sequence are within ε *of each other*. For real numbers, this is equivalent to having a limit.

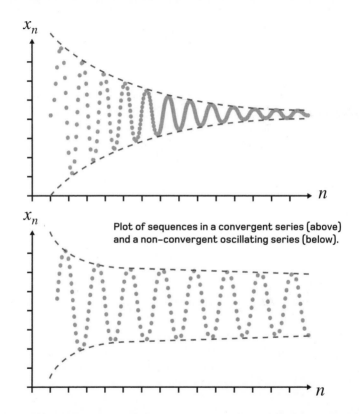

Plot of sequences in a convergent series (above) and a non-convergent oscillating series (below).

Convergent series

The sum of an ordered list of numbers is convergent if it tends towards a specific value or limit. Intuitively, we might imagine that a series settles down if the difference between successive partial sums, the series totalled to a specific number of terms, gets smaller and smaller. For example, if the sequence of partial sums is $(1, S_1, S_2, S_3, \dots)$, where

$$S_n = 1 + \tfrac{1}{2} + \tfrac{1}{3} + \dots + \tfrac{1}{n}.$$

then the difference between S_n and S_{n+1} is $\frac{1}{n+1}$. As n gets very large $\frac{1}{n+1}$ gets very small. But is this really enough to say that this series, known as the *harmonic series* (see page 102), actually settles down to a limit?

It turns out that S_n in this case does *not* settle down, and the series is divergent. So, although successive differences may get small, as with a convergent Cauchy sequence, this on its own is not enough to guarantee that a series converges.

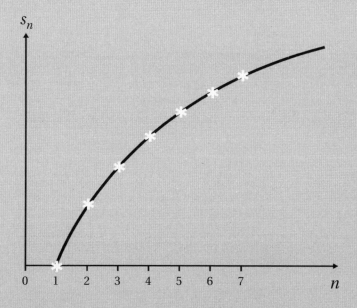

Graph of the harmonic series – although the sums get
gradually closer together, they never converge on a limit.

Estimating π

Many methods for estimating the irrational constant π rely on a sequence approach. As far back as the third century BC, the Greek mathematician Archimedes of Syracuse used a sequence of approximations to find π to two decimal places.

Consider a circle of radius 1, and hence circumference exactly 2π. Sketch a series of regular n-sided polygons within it, starting with a square. Each n-gon can be thought of as a group of triangles with an apex angle $\theta = \frac{360°}{n}$. Dividing each of these in half creates right-angled triangles of hypotenuse length 1, a radius, and one angle of $\frac{\theta}{2}$. Using trigonometric functions (see page 132), we can calculate the other sides of the triangle and hence the perimeter of a polygon.

Of course, Archimedes did not have access to values of the trigonometric functions, so he had to chose n carefully. Modern approaches use series approximations. Isaac Newton expended a lot of time and effort calculating π to 15 decimal places.

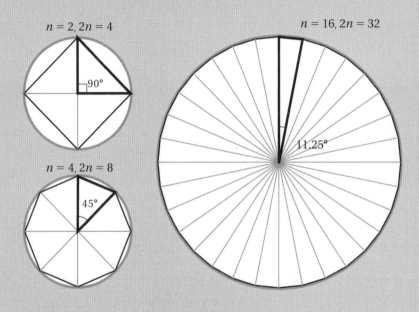

$n = 2, 2n = 4$

$90°$

$n = 4, 2n = 8$

$45°$

$n = 16, 2n = 32$

$11.25°$

Steps in Archimedes' sequence method for
estimating π. Increasing values of *n* give
increasingly accurate estimates of π.

Estimating *e*

Euler's constant, the irrational number *e*, has its origins in the study of sequences, and can be estimated using them. An early encounter with the constant is in the treatment of a problem of compound interest by Jacob Bernoulli in the late 17th century. In compound interest, both the amount invested and the interest accrued by a given time are used to determine the interest paid at the next step. If the interest rate is 100 per cent per year, with half-yearly payments, then for an investment of £1, interest of 50p would be paid after six months, making a new total of £1.50. After another six months, a further 75p would be paid, making £2.25 in total. More generally, for a year divided into *n* equal time periods, our overall return is given by:

$$\left(1 + \frac{1}{n}\right)^n.$$

Bernoulli noted that, as *n* gets larger, this expression converges on the value that we now call Euler's constant: approximately 2.71828182846.

$$e = \lim_{n \to \infty} \left(1 + \frac{1}{n}\right)^n$$

Iteration

Iteration is a mathematical process in which a rule, action or instruction is repeated. This repetition can generate a sequence. Iterative methods are often used in numerical analysis, the study of methods for translating mathematical problems into a language understood by computers.

The subjects of dynamical systems and chaos describe how states of a system evolve when simple rules are applied iteratively. In all these application, it is important to

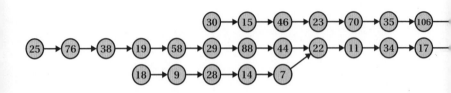

understand the extent to which different initial values can affect the end result. This is not always an easy task.

For example, take a positive integer x. If it is odd, multiply it by 3 and add 1. If it is even, divide it by 2. Now apply the rule again, and only stop applying it if the sequence reaches 1. Every initial value of x that has been tested stops in a finite amount of time. In 1937, German mathematician Lothar Collatz conjectured that this holds true for every possible value of x, but this has yet to be proved.

A map of the *Collatz orbits* by which numbers up to 30 approach the end of the sequence at 1. The number 27 is omitted for practical reasons – it takes 95 additional steps to join this map, connecting to it at the number 46.

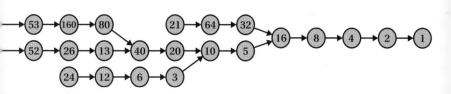

Arithmetic progressions

An arithmetic progression is an ordered list of numbers in which the difference between successive terms is a constant. An example is 0, 13, 26, 39, 52, ..., where the constant common difference is 13. If this common difference is positive, a sequence like this will tend to infinity. If the common term is negative, the sequence tends to negative infinity. The recently proved Green-Tao theorem (see page 316) describes the prevalence of long arithmetic progressions of prime numbers.

Partial sums of an arithmetic progression are relatively simple to calculate using a little trick. For instance, what is the sum of the numbers 1 to 100? The easy way to do this is to list the sum twice, once forwards and once backwards, making columns that sum to 101. Since there are 100 of these, the total sum is 100 multiplied by 101, divided by 2. In general, this argument shows that the sum of any arithmetic progression is given by:

$$a + 2a + 3a + ... + na = \tfrac{1}{2}an(n + 1).$$

$$1 + 2 + 3 + \ldots + 98 + 99 + 100$$

$$100 + 99 + 98 + \ldots + 3 + 2 + 1$$

$$\overline{101 + 101 + 101 + \ldots + 101 + 101 + 101}$$

Geometric progressions

A geometric progression is an ordered list of numbers in which each successive term is the product of the previous term and a constant number. An example is 1, 4, 16, 64, 256, ... where the constant multiplying factor, known as the common ratio r, is 4.

The partial sum of a geometric progression is $S_n = a + ar + ar^2 + ... + ar^n$. If the modulus of r is greater than 1 then this diverges to plus or minus infinity, but if the modulus of r is less than 1, then the limiting series, called a geometric series, tends to the limit $S = \frac{a}{(1-r)}$.

Geometric progressions arise in many mathematical problems, and are fundamental to the study of compound interest and value in accountancy. Many mathematicians would argue that they also resolve Zeno's paradox (see page 84), since the sums of the distance covered and time taken by the hare are geometric progressions that sum to the distance of the race.

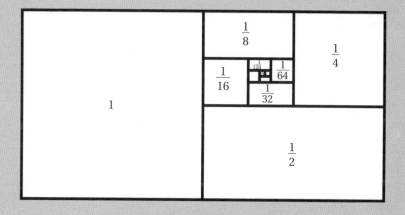

In the diagram above, the areas of the rectangles represent the geometric progression with a common ratio of 1/2, showing clearly that the infinite series converges on a value of 2.

The harmonic series

The harmonic series is the sum of an infinite sequence of steadily diminishing fractions. Important to the theory of music, it is defined as:

$$\sum_{n=1}^{\infty} \frac{1}{n}, \text{ and its first terms are: } 1 + \frac{1}{2} + \frac{1}{3} + \frac{1}{4} + \frac{1}{5} + \cdots$$

One surprising aspect of the harmonic series is that it grows without bounds even though the successive differences between its terms shrink towards zero.

One way of spotting this divergent behaviour is to collect the terms together in smaller groups. This reveals that it is always possible to form a group of successively smaller terms that, together, sum to a number larger than one half. For instance, $\left(\frac{1}{3} + \frac{1}{4}\right)$ is greater than one half, as is $\left(\frac{1}{5} + \frac{1}{6} + \frac{1}{7} + \frac{1}{8}\right)$.

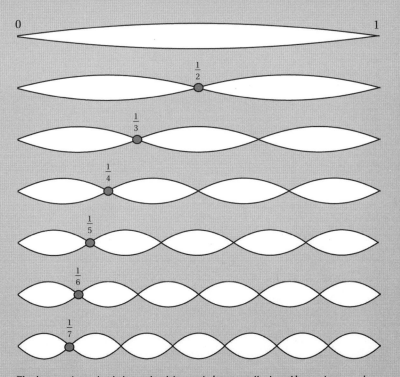

The harmonic series is important in music because it gives the various *modes of vibration* for a plucked or struck string that is fixed at both ends.

Series and approximations

Some of the fundamental numbers of mathematics arise as infinite sums, and so these series can be used to find approximations to numbers, such as π, e and some natural logarithms.

The harmonic series, $1 + \frac{1}{2} + \frac{1}{3} + \frac{1}{4} + \frac{1}{5} + \cdots$, is a good place to start. By changing every other plus sign to a minus, the sum converges on the value of the natural logarithm of 2. And by replacing the denominator of each fraction with its square, the sum converges on the number $\frac{\pi^2}{6}$. In fact, every sum of even powers converges on a known constant multiplied by a power of π^2. The sums of odd powers also converge, but to numbers without a known *closed-form* expression.

Finally, if we replace each denominator with its factorial, the sum converges on e. A factorial, represented by the symbol !, is the product of a number multiplied by all the positive numbers below it. So $3! = 3 \times 2 \times 1 = 6$ and $5! = 5 \times 4 \times 3 \times 2 \times 1 = 120$.

$$1 - \frac{1}{2} + \frac{1}{3} - \frac{1}{4} + \frac{1}{5} - \frac{1}{6} + \frac{1}{7} - \ldots = \ln 2$$

$$1 + \frac{1}{2^2} + \frac{1}{3^2} - \frac{1}{4^2} + \frac{1}{5^2} + \frac{1}{6^2} + \frac{1}{7^2} + \ldots = \frac{\pi^2}{6}$$

$$1 + \frac{1}{2^4} + \frac{1}{3^4} + \frac{1}{4^4} + \frac{1}{5^4} + \ldots = \frac{\pi^4}{90}$$

$$1 + 1 + \frac{1}{2!} + \frac{1}{3!} + \frac{1}{4!} + \frac{1}{5!} + \frac{1}{6!} + \frac{1}{7!} + \ldots = e$$

$$1 + \frac{1}{2 \times 1} + \frac{1}{3 \times 2} + \frac{1}{4 \times 3} + \frac{1}{5 \times 4} + \ldots = 2$$

Power series

A power series is the sum of the terms in an ordered list, where those terms involve increasing positive powers of a variable x. The geometric progression

$$1 + x + x^2 + x^3 + x^4 + \dots$$

is a special case, in which the coefficients of each term are equal to 1. Power series are much more general than they might appear, and many functions can be written as power series. If all the coefficients beyond a given term are zero, then the power series is finite and forms a polynomial (see page 184).

Can power series converge? Using the theory of geometric progressions (see page 100), we can tell that, if x is between -1 and 1, then the partial sum for the series above converges to $\frac{1}{(1-x)}$. Of course not all power series obey such rules, but comparisons with simple geometric progressions can often be used to determine whether or not they do.

$$f(x) = \sum_{n=0}^{\infty} a_n (x - c)^n =$$

$$a_0 + a_1(x - c)^1 + a_2(x - c)^2 + a_3(x - c)^3 + \ldots$$

$$f(x) = \sum_{n=0}^{\infty} a_n x^n =$$

$$a_0 + a_1 x + a_2 x^2 + a_3 x^3 + \ldots$$

Introducing geometry

Geometry is the study of shape, size, position and space. In the classical form established by Greek mathematician Euclid around 300 BC, it is based on lists of objects, and assumptions called axioms, from which all results follow.

Euclid's influential book *Elements* listed five axioms:
1. *A line can be drawn between any two points.*
2. *A line segment can be extended infinitely in either direction.*
3. *A circle can be drawn of any radius centred at any point.*
4. *Any two right angles are equal.*
5. *For a given straight line and a point not on the line, there is precisely one line through the point, the parallel line, that does not intersect the original line.*

It is worth noting that Euclid's axioms use a number of terms, such as line, right angle and radius, without explanation or definition. As a result, new axioms were introduced in the late 1800s to develop geometry within a strictly logical framework.

Lines and angles

Lines and angles are two of the most fundamental terms in geometry. Euclid's fifth axiom says that, given a straight line and a point not on that line, all but one of the possible lines through the point will inevitably intersect the given line. In other words, typical lines intersect, and non-intersecting parallel lines are unusual.

The concept of angle originated as a tool to describe how lines intersect. Suppose two lines intersect at a point P, as shown opposite. In this case, a circle centred on P is divided into four segments by the lines. If these segments are of equal area, then the lines are said to be perpendicular and the angles are right angles. This relates to Euclid's fourth axiom.

In more general cases, angles are measured in degrees. Through trigonometric functions (see page 132), angles also play a fundamental role in areas seemingly unconnected with geometry.

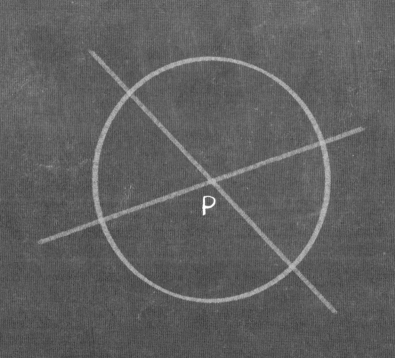

Measuring angles

Historically, measuring angles between two lines involved drawing a circle around their point of intersection and dividing it into a number of equal segments or units. Ancient Mesopotamian astronomers introduced the idea of using 360 such divisions, which we know today as degrees. They also subdivided the degree units into 60 equal minutes, each containing 60 equal seconds. To avoid confusion with units of time, these smaller subdivisions are often known as minutes of arc and seconds of arc. Thus the measure of an angle is obtained by working out how many degrees, minutes and seconds make up the angle.

The numbers 60 and 360 are very convenient to use in this context, since 60 can be divided by 1, 2, 3, 4, 5 or 6 and still result in a whole number. However, the specific units are not essential to angle measurement. The fundamental idea here is that we can think of an angle as being the proportion of the circle that is enclosed by the two lines forming the angle.

Circles

A circle is defined as the set of points that lie at an equal distance, the radius r, from a central point P, and is one of the primitive elements taken for granted in Euclid's axioms. The closed curve through all the outer points is the circle's circumference, and the length of circumference C is linked to the radius r by the equation $C = 2\pi r$, while the area A of the circle is defined by another equation, $A = \pi r^2$. In this way, the circle inevitably leads to one of the two great constants of mathematics, π (see page 40).

The circle also defines other curves, lines and areas. An arc is a limited part of the circumference, while a sector is a region of the circle bounded by two radii and an arc. A chord is a straight line across the circle between two points on its circumference, and a segment is an area within the circle bounded by the circumference and a chord. A secant is an extended chord – a straight line cutting the circle at two points – while a tangent is a straight line that touches the circle at a single point.

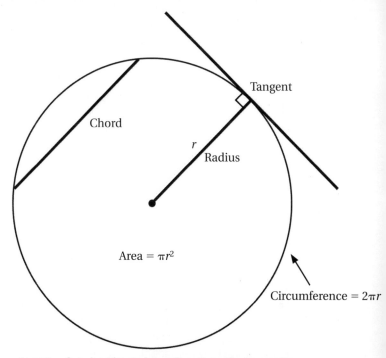

Aspects of circles. The circle's radius, circumference and area
are intimately linked to the definition of the constant π, while
various geometric lines and areas are also derived from circles.

Radian angles

As an alternative to the traditional degrees, minutes and seconds of arc, mathematicians often express angles in units called radians. Based on the geometry of a circle, radians have many advantages. In particular, they make the handling of trigonometric functions (see page 132) far easier.

The intuitive meaning of radians is best understood by considering a circle of radius 1. The angle in radians between two lines then equals the length of the arc between the two lines formed by the circle of radius 1, centred on the intersection of our two lines.

Since the circumference of a circle is given by $C = 2\pi r$, if $r = 1$, then $C = 2\pi$. Therefore a portion x of the circle has angle θ radians, where $\theta = 2\pi x$. For example, cutting the circle into four equal segments gives a right angle, which is equal to 2π multiplied by $\frac{1}{4}$, or $\frac{\pi}{2}$ radians.

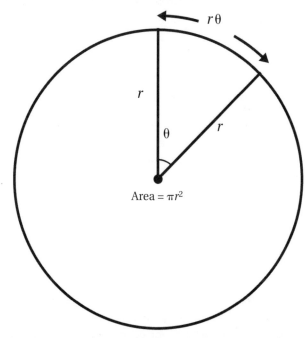

$r\theta$

r

r

θ

Area $= \pi r^2$

A neat consequence: If a sector of a circle of radius r, a piece of cake say, has an angle of θ radians, then the length of the circular arc of the cake is simply $r\theta$, so angles measured in radians offer a simple way of measuring arc lengths.

Triangles

A triangle can be defined by any three points that do not lie along the same straight line. The triangle is simply the region enclosed by the three line segments connecting those points.

The area of a triangle can be calculated by constructing rectangles around it. If we choose one side to be the base of the triangle and define the altitude to be the perpendicular distance of the third vertex of the triangle to the base, then the area of the triangle is half the product of the altitude with the length of the base.

Triangles and their higher-dimensional generalizations are often used as simple ways of describing more complicated bodies. For example many objects can be modelled by gluing triangles together. This idea is of course familiar to engineers, who break complicated shapes such as curving walls into straight-sided triangles to give them greater strength.

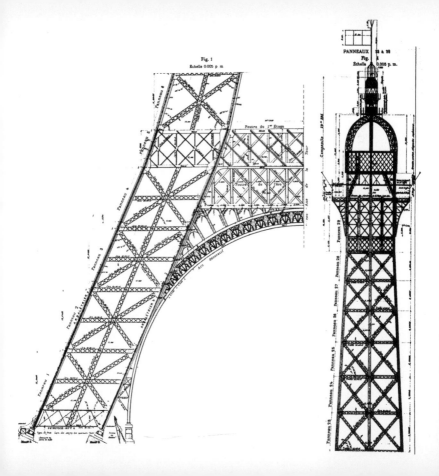

Fig. 1
Échelle 0.005 p m

Types of triangle

There are several special types of triangle, each given their own specific names. In every triangle, the sum of the internal angles is π radians (or $180°$), and there is a clear relationship between the size of the angles and the relative lengths of the sides.

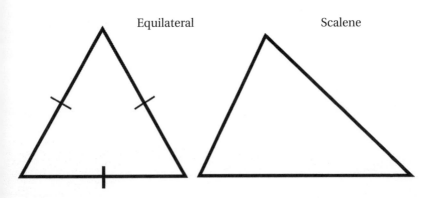

Equilateral

Scalene

An equilateral triangle has all three sides equal, which also means all three angles are equal. Since the angles sum to π radians, they must each equal $\frac{\pi}{3}$, or 60°. An isosceles triangle has two sides equal, and must therefore have two angles equal.

A right-angled triangle has one angle that is a right angle, $\frac{\pi}{2}$ or 90°, and a scalene triangle has three sides of different lengths and three angles of different sizes.

Isosceles

Right-angled

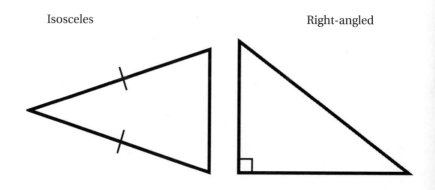

The centre of a triangle

There are many different ways to define the centre of a triangle. For example, it could be a point equidistant from all three vertices, the centre of the largest circle that can be drawn inside the triange or the centre of a circle that touches each of its three corners. These are all *natural* definitions, though they may not coincide in the same positions.

One of the most useful triangle centres is the centroid. If you draw a line from each corner of a triangle to the mid-point of the opposite side, then the centroid is where the three lines meet. The fact that these three lines *do* meet at a single point is not completely obvious. The centroid marks a point that would be the centre of mass if the triangle were cut out from a material of uniform density. If you hung such a triangle from any other point, it would find an equilibrium position, with the centroid below the pivot point, on a vertical line through the pivot.

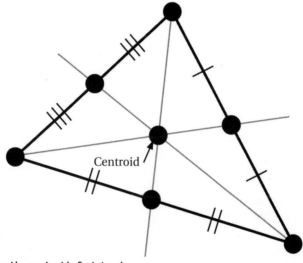

Centroid

Finding the centroid of a triangle

Polygons

In essence, a polygon is simply a closed region bounded by a number of straight lines. However, the term is often used as shorthand for a specific type of polygon, the *regular* polygons whose sides are all of equal length. These include pentagons, hexagons, heptagons, octagons and so on.

Regular polygons can be constructed using triangles that have two matching angles, known as isosceles triangles. The peaks of each triangle meet at the centre of the new shape, as shown. Since the sum of their central intersections must be 2π radians, the angles at each peak equal $\frac{2\pi}{n}$, where n is the number of triangles, or sides of the polygon. Because we know that the three angles in a triangle sum to π radians we know that the sum of the equal angles $2a$ is given by $2a = \pi - \left(\frac{2\pi}{n}\right)$. The quantity $2a$ is also the value of each of the internal angles of the regular polygon. For instance, in a pentagon with $n = 5$, the internal angles are $\frac{3\pi}{5}$.

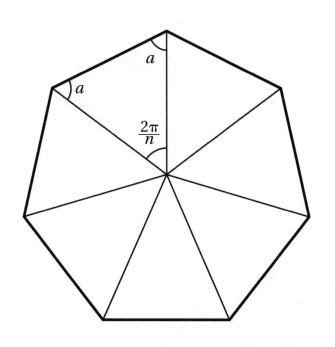

Similarity

Two objects are said to be *similar* if they are rescaled versions of each other. This is one of the many ways of saying that two objects have the same shape. In the case of triangles, if the three angles in one triangle equal those of another, they are similar. Equivalently, this means that the ratio of the lengths of two sides is the same in both triangles.

When considering other geometric objects, such as polygons and curves, there are other criteria to be met. For instance, two regular polygons are said to be similar if they have a matching number of sides.

The term similarity, or *similarity transformation*, is also used to describe a scaling operation by which an object is transformed into a similar object. Similarity transformations multiply the Cartesian coordinates (see page 160) of all points in Euclidean space by the same factor, decreasing or increasing the size of an object without altering its shape.

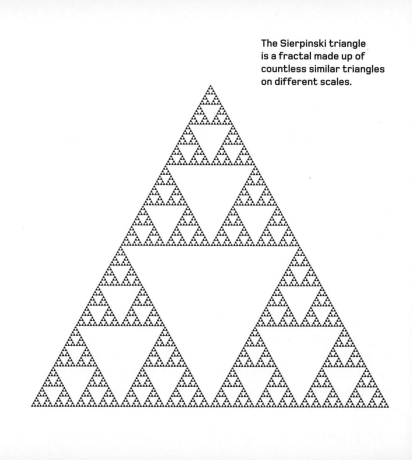

The Sierpinski triangle is a fractal made up of countless similar triangles on different scales.

Congruence

Two objects are said to be *congruent* if they are the same shape and size. So two triangles are congruent if they are similar – same shape – and the lengths of corresponding side are equal – same size. The *scaling factor* between them is 1.

Note that congruence does not necessarily mean that one triangle can be moved over the other to match perfectly, simply by translations in the plane. Two congruent triangles may be mirror-images of each other, which can only be matched up physically by lifting one of them out of the plane entirely.

Two general triangles are congruent if any of the following three sets of quantities are the same: the lengths of the three sides; the lengths of two sides and the angle between them; or the length of one side and the angles made with the other sides at either end. Thus any of these three criteria is enough to specify a triangle.

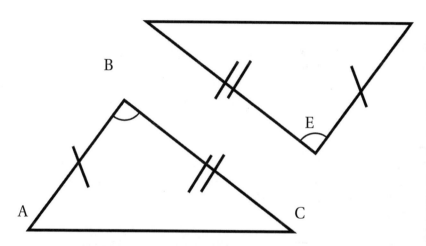

A pair of congruent triangles can be recognized by various tests, such as having at least two matching sides and one matching angle as shown here. Despite this, these two congruent triangles cannot be overlaid on each other.

Pythagoras' theorem

Although it is named after the Greek mathematician Pythagoras from the late sixth century BC, this famous connection between the lengths of the sides of a right-angled triangle was almost certainly known to the Babylonians many centuries earlier.

It states that the square of the longest side, known as the hypotenuse, is equal to the sum of the squares of the lengths of the other two sides. A simple proof based on the ratios of the sides of similar triangles is illustrated opposite, but the theorem can also be proved by considering the areas of geometric squares constructed on each side of the triangle.

Pythagoras' theorem is an important tool of geometry, and many definitions of distance in coordinate geometry (see page 160) are based on this relation. It can be rephrased in terms of the relationship between the trigonometric functions, sine and cosine (see page 136).

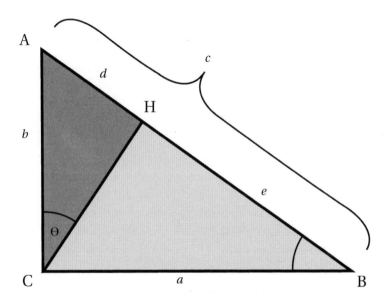

Similarity of the triangles **ABC** and **CHB** on the one hand, and **ABC** and **CAH** on the other, implies that: $\frac{a}{c} = \frac{e}{a}$ and $\frac{b}{c} = \frac{d}{b}$.

Hence: $a^2 = ec$ and $b^2 = dc$, or: $a^2 + b^2 = (e + d)c = c^2$.

Sine, cosine and tangent

Right-angled triangles allow us to associate functions with angles, via the ratios of the lengths of the sides. These are called *trigonometric functions*, and the fundamental functions defined in this way are the sine, cosine and tangent functions.

To define these functions, choose one of the angles, θ, which is not 90°. It is formed by the intersection of the hypotenuse of length H and another side, called the adjacent side, of length A. The remaining side opposite the angle has length O. The sine, cosine and tangent functions are then defined by the ratios:

$$\sin \theta = \frac{O}{H} \; ; \quad \cos \theta = \frac{A}{H} \; ; \quad \tan \theta = \frac{O}{A}$$

Since any two right-angled triangles with an angle θ are rescaled versions of each other, the functions return the same answer regardless of the size of the triangle. What's more, since $\frac{O}{A} = \frac{O}{H} / \frac{A}{H}$, we can see that $\tan \theta = \frac{\sin \theta}{\cos \theta}$.

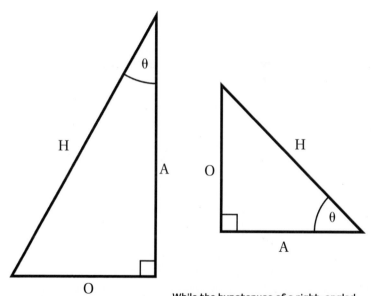

While the hypotenuse of a right-angled triangle is always the longest side, the opposite and adjacent sides are defined in relation to the angle being considered.

Triangulation

Triangulation is a method for calculating the complete properties of a triangle from the measurement of just one side and an angle. It relies on knowing the values of the trigonometric functions, sine, cosine and tangent.

Imagine a prince trying to reach Rapunzel at the top of her doorless tower. How can he know the height d of her window, and how long Rapunzel's hair must be to reach the ground? He stands at a distance l from the tower, and measures the angle θ between the base of the tower and the window.

Assuming the tower is vertical, its window and base, along with the prince's location, form the corners of a right-angled triangle. The prince knows angle θ and the adjacent side l, and wants to find d, the opposite side to angle θ. Plugging these values into the tangent formula, we can see that:

$$\tan \theta = \frac{d}{l} \, , \ \text{ and so } \ d = l \times \tan \theta.$$

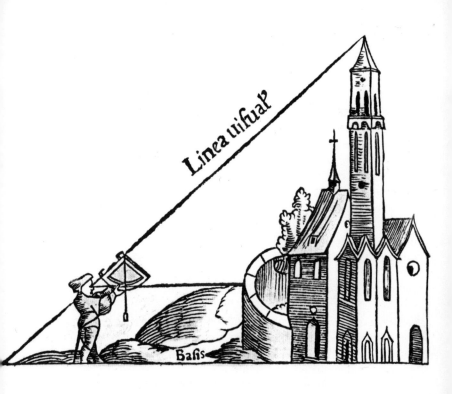

Trigonometric identities

Trigonometric identities are expressions involving the functions sine, cosine and tangent that are true for all angles. Given any right-angled triangle with angle θ, opposite length O, adjacent length A and hypotenuse H, Pythagoras' theorem states that $O^2 + A^2 = H^2$. Dividing both sides of this equation by H^2 gives:

$$\frac{O^2}{H^2} + \frac{A^2}{H^2} = 1, \quad \text{or} \quad \left(\frac{O}{H}\right)^2 + \left(\frac{A}{H}\right)^2 = 1$$

And since $\sin\theta = \frac{O}{H}$ and $\cos a = \frac{A}{H}$ this means that:

$$\sin^2\theta + \cos^2\theta = 1$$

for any angle θ. Note that the form $\sin^2\theta$ shows that we are talking about *the square of the sine of* θ, rather than *the sine of* θ^2. This identity is true for all values of θ, but it tells us something useful about the functions themselves. Note that it is effectively a restatement of Pythagoras' theorem.

For a right-angled triangle in which the length of the hypotenuse H and the angle *a* are known, the definitions of sine and cosine make it easy to find the lengths of the other sides.

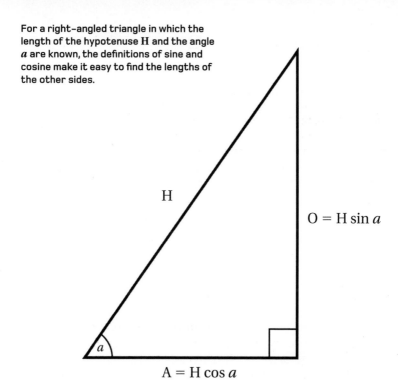

H

$O = H \sin a$

$A = H \cos a$

Sine and cosine rules

The sine and cosine rules are formulae relating the angles and sides of general triangles. The idea of congruence (see page 128) shows that two sides and the angle between them determine a triangle, so it should be possible to find the other angles and the other side from this information.

For a triangle with sides and angles as shown opposite, the rules are:

$$\frac{\sin A}{a} = \frac{\sin B}{b} = \frac{\sin C}{c} \text{ (sine rule)}$$

$$c^2 = a^2 + b^2 - 2ab \cos C \text{ (cosine rule)}$$

If C is a right angle, then $\cos C = 0$ and the cosine rule is just Pythagoras' theorem. We can therefore think of the cosine rule as giving a correction to Pythagoras' theorem for cases where C is not a right angle.

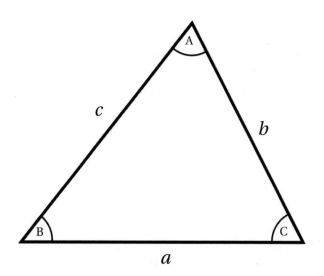

Double angle formulae

The double-angle formulae allow us to work out the sines and cosines of sums of angles. They also allow the usefulness of sines and cosines to be extended beyond the narrow range of angles $(0-90°)$ permitted in a triangle.

The formulae are derived by considering triangles made up of two triangles glued together, as illustrated opposite:

$$\sin (A + B) = \sin A \cos B + \cos A \sin B$$

$$\cos (A + B) = \cos A \cos B - \sin A \sin B$$

Setting $A = B$ gives the generalized double angle formulae:

$$\sin (2A) = 2 \sin A \cos A$$

$$\cos (2A) = \cos^2 A - \sin^2 A = 1 - 2 \sin^2 A$$
$$= 2 \cos^2 A - 1$$

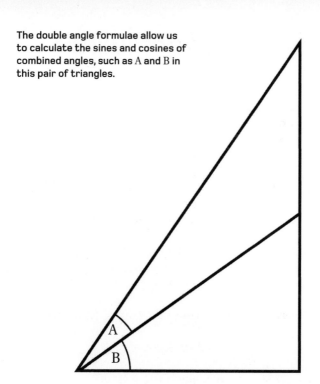

The double angle formulae allow us to calculate the sines and cosines of combined angles, such as A and B in this pair of triangles.

Introducing symmetry

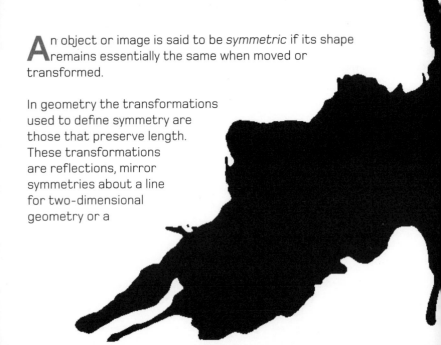

An object or image is said to be *symmetric* if its shape remains essentially the same when moved or transformed.

In geometry the transformations used to define symmetry are those that preserve length. These transformations are reflections, mirror symmetries about a line for two-dimensional geometry or a plane for three-dimensions; rotations, in which the object is moved around a plane or rotated around an axis; and translations, in which the object is moved in a given direction. These actions can also be combined. If the application of a given transformation to an object does not appear to change it, the object is said to be *invariant* under the transformation.

Symmetry is also useful in other areas of mathematics, where any operation on a mathematical object can be considered symmetric if it preserves some property of that object. This is an important concept used in the definition of groups of operations (see page 268).

Translation, rotation and reflection

There are three basic types of symmetry within geometry. These are the ways in which we can transform an object while preserving its essential shape.

Translations move the shape in a given direction, but do not change the lengths or angles that specify the object. Rotations spin the shape around some point in the plane, again without changing the lengths or angles involved.

In two dimensions, reflections mirror the shape across any given line, known as the *axis of symmetry*. While other translations can be achieved by sliding a shape around in its plane, reflection can only be achieved by lifting a shape out of the plane and turning it over. Again, the lengths or angles are unchanged. In some circumstances the inclusion of reflections in the definition of a symmetry might be inappropriate. For example, the two sides of a jigsaw piece are not equivalent, since one side includes the picture, and the other is blank.

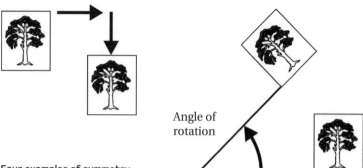

Angle of rotation

Four examples of symmetry transformations. Above: translation and rotation. Below: Reflection, and a glide symmetry consisting of a reflection, here in a horizontal line, and a translation.

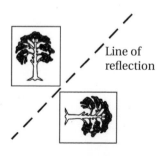

Line of reflection

Polyhedra

The three-dimensional equivalent of a polygon is a polyhedron, a volume whose boundary is made up of flat, two-dimensional faces. Just as there are particularly simple regular polygons to which certain rules apply, there is also a family of five regular polyhedra, known as the Platonic solids:

- *Tetrahedron: four faces, each of which is an equilateral triangle.*
- *Cube: six faces, each of which is a square.*
- *Octahedron: eight faces, each of which is an equilateral triangle.*
- *Dodecahedron: twelve faces, each of which is a regular pentagon.*
- *Icosahedron: twenty faces, each of which is an equilateral triangle.*

Of course, with so much more freedom to place faces, there are many more types of polyhedron than there are polygons.

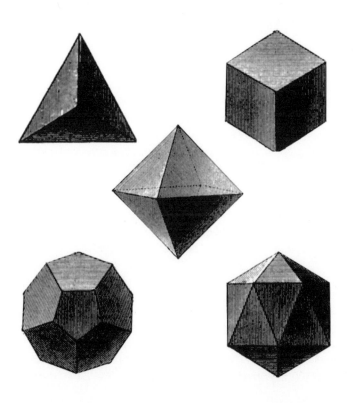

Tessellations

Two-dimensional shapes are said to *tessellate a region* if they can fit together side-by-side with no gaps or overlaps to cover the region. Of the regular polygons, only the four-sided square and six-sided hexagon can tessellate the entire plane on their own.

More complicated tessellations of the plane can be constructed by using combinations of shapes. The simplest, known as *periodic tilings*, have translational symmetry. This means the pattern can be shifted in a given direction so that it fits precisely over itself.

Amongst the regular polyhedra, only the cube can tessellate three-dimensional space, but by using more complicated polyhedra, it is possible to obtain infinitely many tessellations called honeycombs. These are important in crystal chemistry, where vertices of the polyhedra mark the positions of atoms in the crystal. Analysis of honeycombs reveals 230 independent tessellations, limiting the range of possible crystal structures.

Penrose tilings

Penrose tilings are a special class of tilings using two different basic shapes. Discovered in the mid-1970s by British theoretical physicist Roger Penrose, these *aperiodic tilings* do not repeat in a periodic pattern.

Remarkably, these abstract objects proved to have a natural application. In the early 1980s, materials scientists discovered aperiodic structures called quasicrystals, with a similar mathematical description. These can be used as hard coatings for other materials, and have very low friction.

The simplest Penrose tilings are constructed using a 'fat' rhombus and a 'thin' rhombus as basic shapes, as shown opposite. A rhombus is a shape with four equal sides, in which each pair of opposite sides are parallel. It is not known whether it is possible to find a single shape that can be assembled with the same properties.

Spheres

A sphere is the three-dimensional equivalent of a circle, a perfectly round geometric object. If the sphere has a fixed frame of reference, for example the polar axis of the Earth, then any location on its surface can be described by two angles. In the Earth's case, we represent these as longitude and latitude. Latitude is the angle between a line joining the location to the sphere's centre, known as a ray, and the main axis. Longitude is the angle around the axis, between the latitude ray and a line from a defined reference point, such as the Earth's prime meridian.

The rays from the boundary of any area on the surface of the sphere form a generalized cone at the centre. The spread of this, called its solid angle, is a measure of the proportion of the area of intersection of this cone with the overall surface area of a sphere of radius 1. Since the surface area of a sphere is given by the formula $4\pi r^2$, the surface area of this sphere is simply 4π.

Representing the surface of a spherical object on a flat piece of paper requires choices to be made: should the picture be such that the ratio of two areas is the same, should lines of latitude be straight, or should some other measure be preserved? This leads to different two-dimensional representations of the same curved surface.

Non–Euclidean and non-classical geometries

A non-Euclidean geometry is one based on a surface or in a space other than the familiar flat planes of Euclidean geometry (see page 108). In these circumstances, Euclid's fifth axiom, that there is precisely one line through a point parallel to another given line, does not apply. For instance, consider the geometry of a spherical surface. Here, a line translates into an arc of a great circle around the circumference of the sphere. If we choose a point that is not on this line, then any other great circle through the new point will intersect with our original circle. So there are no parallel lines on the surface of a sphere!

Non-Euclidean geometries can be divided into elliptic geometries with positive curvature, such as the surface of a sphere, and hyperbolic geometries with negative curvature, such as the *saddle* illustrated opposite. It is also possible to construct so-called *non-classical geometries*, in which there may be many lines through a given point parallel to a given line.

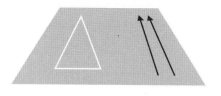

Zero
curvature

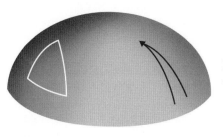

Positive
curvature

Negative
curvature

Sphere-packing problem

The sphere-packing problem is to determine the most efficient arrangement of spheres in a box – i.e. how should the spheres be arranged so as to minimize the amount of unoccupied space?

Despite being most clearly relevant to a greengrocer shipping oranges, this problem has a rich history, with oranges replaced by cannonballs. The 17th-century German astronomer and theoretician Johannes Kepler conjectured that the simple configuration obtained by starting with a square horizontal array of spheres and then placing another layer in the gaps created by these, and so on, is best. Kepler calculated that this occupies a little more than 74 per cent of the available space – the same as a related hexagonal arrangement.

That these two are indeed the best orderings has been very hard to prove. An exhaustive proof using computers to analyse the many differnet special cases was completed in 2003.

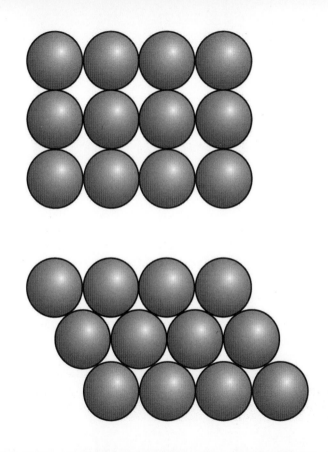

Conic sections

Conic sections, like lines and planes, are fundamental to Greek geometry. They arise from taking slices through a three-dimensional cone, creating a set of geometrically beautiful curves.

If the axis of the cone is vertical with a tip at O, then:
- *A circle is formed by the intersection of the cone with any horizontal plane not passing through O.*
- *A parabola is created by the cone's intersection with a plane parallel to the cone that does not pass through O.*
- *An ellipse is created by the cone's intersection with a non-horizontal plane not passing through O, if the angle of the plane is greater than the angle of the cone.*
- *A pair of hyperbolas is formed as in the previous case, if the angle of the plane is less than the angle of the cone.*

The special case of planes through O give a single point, or one or two straight lines.

Parabola

Circle

Ellipse

O

O

Hyperbola

Cartesian coordinates

Cartesian coordinates describe the position of a point on a plane, using a pair of numbers that describe how to reach that point from an arbitrary origin. Introduced in the 19th century, by the French philosopher and mathematician René Descartes, they work rather like the coordinate systems used on maps and make it easier to talk about geometric objects.

In the two-dimensional plane, a point has coordinates (x, y), meaning you should move x units in the horizontal direction, and then y units in the vertical. Negative points like $(-1, -2)$ involve moving in the opposite directions.

Similarly in three dimensions, three coordinates (x, y, z) are used to specify points. It's easy to see how this makes it possible for mathematicians to talk easily about n-dimensional spaces specified by n coordinates, even if such multidimensional space is hard for us to visualize.

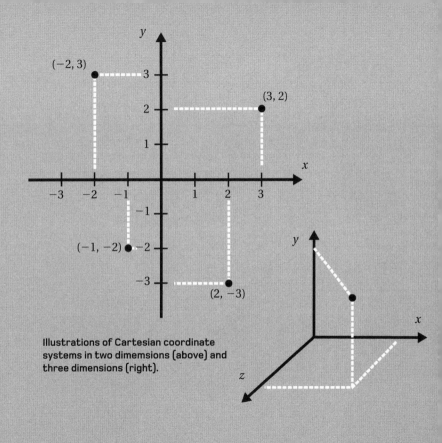

Illustrations of Cartesian coordinate systems in two dimensions (above) and three dimensions (right).

Algebra

Elementary algebra is the art of manipulating mathematical expressions, with quantities represented by symbols, whilst abstract algebra is the theory of mathematical structures such as groups (see page 268). The use of symbols rather than numbers makes it possible to work more generally, and x is the most traditional option when it comes to representing an unknown number or arbitrary number. Using this approach we can manipulate expressions and rewrite relationships between quantities in different, more concise, ways.

Suppose, for example, we are asked to find the number which, when added to 3, gives a total of 26. Of course we can probably work this out instinctively, but mathematically we can use a letter to represent the unknown, and express the puzzle in the equation $x + 3 = 26$. In this trivial example, we can see that subtracting three from both sides will give us an answer, in the form $x = 26 - 3$. Algebra is all about this kind of manipulation, although the process is usually a little more complex.

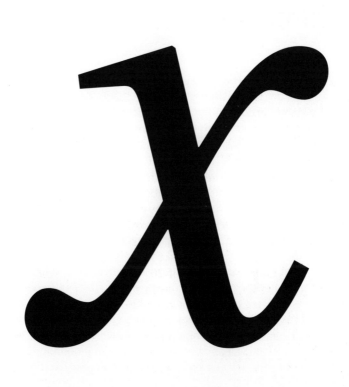

Equations

An equation is a mathematical expression that states that one thing is equal to another. So, $2 + 2 = 4$ is an equation, and so is $E = mc^2$, or $x + 3 = 26$. Each of these examples is subtly different. The first is an *identity* – it is always true. The second is a *relationship*, defining E in terms of m and c, whilst the third is an equation which is only true for certain values of x. In most algebraic contexts, at least one side of the equation will involve unknown elements, usually denoted by x, y or z. Many algebraic techniques are concerned with manipulating and solving equations to find these unknowns.

Most quantitative disciplines, such as science, economics, and areas of psychology and sociology, describe real-world situations in terms of equations. In physics, for example, Newton's laws of motion, describing the interaction of masses and forces, can be written as equations involving derivatives (see page 208) as well as numbers, and in some economic models, equations link the price of goods to supply and demand.

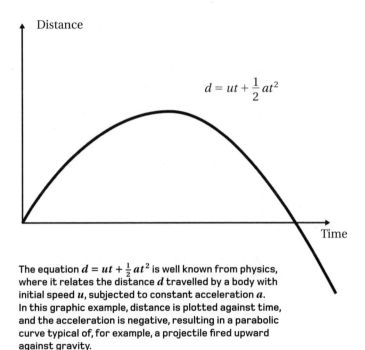

Distance

$$d = ut + \frac{1}{2}at^2$$

Time

The equation $d = ut + \frac{1}{2}at^2$ is well known from physics, where it relates the distance d travelled by a body with initial speed u, subjected to constant acceleration a. In this graphic example, distance is plotted against time, and the acceleration is negative, resulting in a parabolic curve typical of, for example, a projectile fired upward against gravity.

Manipulating equations

Equations can be simplified and in some cases solved by manipulating them in a variety of ways. There are also conventions about how to represent equations. One of the most common is that of missing out multiplication signs, perhaps wise given the ubiquity of x as the all-purpose symbol for unknown variables. So instead of writing $x \times y$, we simply write xy, and $E = mc^2$ means $E = m \times c \times c$. Brackets, meanwhile, are used to clarify potentially confusing expressions.

The expression $2 \times 3 + 5 \times 4$ is ambiguous: the answer depends on the order in which the operations are processed. Brackets are used to indicate the order to be followed: start with the simple expressions nested in the greatest number of brackets, and then work outwards. Thus, $(2 \times 3) + (5 \times 4)$ gives a different result to $2 \times (3 + 5) \times 4$ and $2 \times (3 + (5 \times 4))$. Brackets are not always necessary, for instance in associative operations such as multiplication, where $a \times b \times c$ gives the same result as $a \times (b \times c)$ and $(a \times b) \times c$.

Rules of
Algebraic Manipulation

Subtraction:

If $a + c = b + c$ then $a = b$

Cancellation:

If $ac = bc$ and $c \neq 0$, then $a = b$

Factorization:

$$ab + ac = a(b + c)$$

Simultaneous equations

Simultaneous equations are sets of equations containing several unknowns. An example would be two equations featuring two unknowns, such as: $2x + y = 3$, $x - y = 1$. By solving these two equations together, we can find each of the unknowns.

If we rearrange the second equation, according to the rules of algebraic manipulation, we can express x as $1 + y$. Using this value for x in the first equation, we see that $2(1 + y) + y = 3$, so $2 + 2y + y = 3$, or $2 + 3y = 3$. Rearranging this expression, we see that $3y = 3 - 2$, or $y = \frac{1}{3}$. If we insert this value for y in the second equation we can now work out that x has a value of $\frac{4}{3}$.

In general, one equation is needed for each unknown, though this does not guarantee a solution, nor that a solution is unique. From a geometric point of view, the two equations above are linear: they describe straight lines. So solving a pair of linear equations is the same as finding the intersection of two lines.

Equations and graphs

Plotting an equation as a graph shows the way that the value of one variable changes as the other changes. This uses the idea that any equation relating two real variables can be pictured as a relationship between two-dimensional Cartesian coordinates, x and y. An equation can therefore be interpreted as a curve representing the corresponding values of x and y determined by that equation.

The equation $y = x^2$ generates a parabolic curve of points, as shown. More complicated equations can create more complicated curves, though for each x there may be none or many corresponding values of y.

When a pair of simultaneous equations are plotted on the same axes, the intersections mark points where x and y satisfy both equations. Thus, the solution of simultaneous equations is essentially a question of determining the intersection points of curves: algebra and geometry meet.

Finding the solutions where variables are defined by a pair of equations is essentially the same problem as finding the intersections on a graph

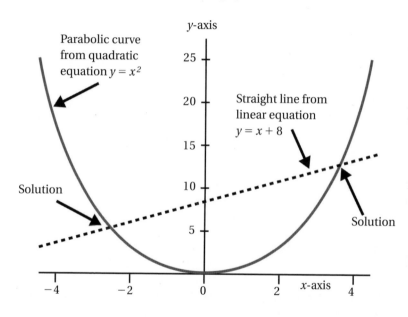

y-axis

Parabolic curve from quadratic equation $y = x^2$

Straight line from linear equation $y = x + 8$

25

20

15

10

5

Solution

Solution

−4   −2   0   2   4   *x*-axis

The equation of a straight line

Any straight line in the plane can be written as either $x = a$, where a is constant (this is the special case of a vertical line) or the more standard form $y = mx + c$, where m and c are constants. The constant m represents the slope of the line and c is the value of y where the line meets the y-axis.

The slope, or gradient, of the line is calculated by considering any two points on the line. It is equal to the change in height between the points, divided by the change in horizontal position between the points. Mathematically, given any two distinct points (x_1, y_1) and (x_2, y_2), then $m = \frac{(y_2 - y_1)}{(x_2 - x_1)}$. The slope of the graph shown opposite is $\frac{4}{5}$.

Both the equations $x = a$ and $y = mx + c$ can be written in the more general form $rx + sy = t$ for suitably chosen constants r, s and t. It is in this form that the equation of a line often appears in simultaneous linear equations (see page 170).

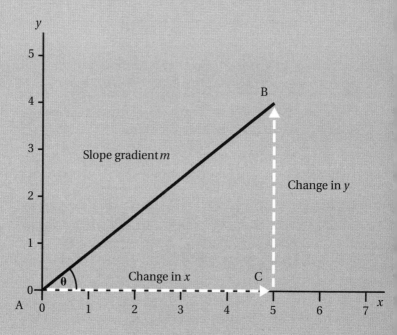

The equation of a plane

A plane is a two-dimensional flat surface in three-dimensional space. The equation of a plane is the three-dimensional generalization of the equation of a line: $ax + by + cz = d$, where a, b, c and d are constants, and at least one of a, b or c is non-zero. Note that, because we are now working in three dimensions, an additional variable z is needed to describe the third direction.

In the special case that $a = b = 0$, the equation reduces to $cz = d$, or $z = \frac{d}{c}$. Since c and d are constants, z is also constant, so this plane is a horizontal surface of constant height z, on which x and y can take any values.

The solution of three simultaneous linear equations in three variables represents the intersection of three planes. It is usually a point, but there can be cases with no solutions (two planes parallel and not coincident) or infinitely many solutions: either a line of solutions or a plane of solutions.

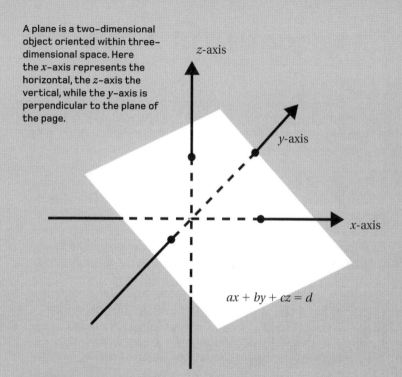

A plane is a two-dimensional object oriented within three-dimensional space. Here the x-axis represents the horizontal, the z-axis the vertical, while the y-axis is perpendicular to the plane of the page.

z-axis

y-axis

x-axis

$ax + by + cz = d$

The equation of a circle

A circle is defined to be the set of points that lie a fixed distance away from a given point. It can also be described in algebraic terms, in the form of an equation.

If the centre of a circle is defined as the origin of a Cartesian coordinate system $(0, 0)$, then we can use Pythagoras' theorem to find the coordinates of an arbitrary point on the perimeter of the circle (x, y). Any radius r linking the centre to the point (x, y) can be treated as the hypotenuse of a triangle, whose other sides have length x and y.

Thus for a fixed radius r, we can write $x^2 + y^2 = r^2$, and define the circle to be the set of points whose coordinates meet that condition. This is the equation of a circle. It provides the starting point for the equations which arise from the various conic sections (see page 158).

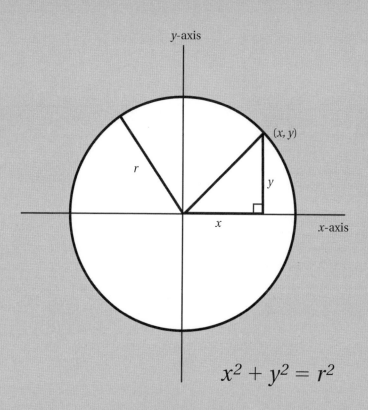

Parabolas

A parabola is one of the conic sections, obtained by intersecting a cone with a plane parallel with the surface of the cone. It has a single maximum or minimum value, and in algebra it is defined by an equation in which one variable is equal to a quadratic function of the other, $y = ax^2 + bx + c$.

The simplest example is $y = x^2$. Since x^2 is greater than zero for both positive and negative values, the smallest value that y can take is zero, when $x = 0$. Moreover as the magnitude of x gets very large, so does x^2.

Parabolas are useful for describing the movement of objects experiencing constant acceleration, since the distance an accelerating object travels is proportional to the square of the time interval involved. For instance, the idealized trajectory of a projectile such as a cannonball has a steady horizontal velocity in the x-direction, but is influenced by acceleration due to gravity acting downwards in the y-direction (see page 165).

Equations of conic sections

A conic section curve is defined geometrically by the intersection of planes with a double-sided cone (see page 158). The algebraic formula for such a cone with symmetry around the z-axis is $|z| = x^2 + y^2$, where $|z|$ is the modulus of z, so $|z|$ equals z if z is positive, and $-z$ if z is negative. The modulus is never negative and measures the size of z.

The z-coordinate of a horizontal plane is a constant, for example c, and its intersection with a vertical cone is defined by $x^2 + y^2 = |c|$. This is equivalent to the equation for a circle of radius $\sqrt{|c|}$. For the intersection with a vertical plane, the y-coordinate is constant, giving $x^2 + c^2 = |z|$. This is the equation for a pair of parabolas, one for $z < 0$ and one for $z > 0$.

The ellipse and hyperbola are produced by intersections with tilted planes. If the plane cuts the cone on just one closed curve, the result is an ellipse of form $\frac{x^2}{a^2} + \frac{y^2}{b^2} = 1$. If it cuts twice, the result is a pair of hyperbolas for which $\frac{x^2}{a^2} - \frac{y^2}{b^2} = 1$.

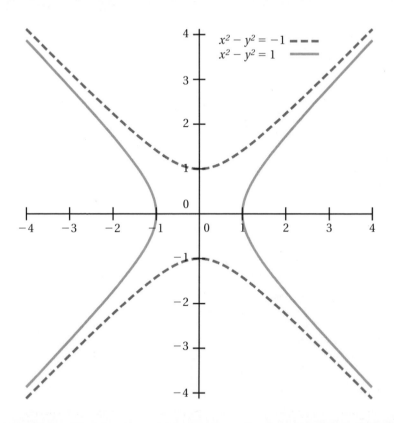

Ellipses

An ellipse is defined as a conic section by the intersection of a tilted plane with a double-sided cone. Such a cone can be defined by the equation $|z| = x^2 + y^2$. If the tilted plane cuts the cone along just one curve, the result is an ellipse of the form $\frac{x^2}{a^2} + \frac{y^2}{b^2} = 1$. The constants a and b are related to the lengths of the shape's long and short axes.

If $a > b > 0$, then the foci of the ellipse are the two points lying on the ellipse's major axis, the x-axis in this case, at a distance $\sqrt{(a^2 - b^2)}$ from the centre. An ellipse can also be defined as the set of points such that the perimeter of the triangles formed by the points on the ellipse and the two foci is constant. In 1609, German astronomer Johannes Kepler observed that planetary orbits can be described by ellipses with the Sun at one focus. In general, ellipses can therefore describe the motion of objects within gravitational fields, such as artificial satellites in their orbits.

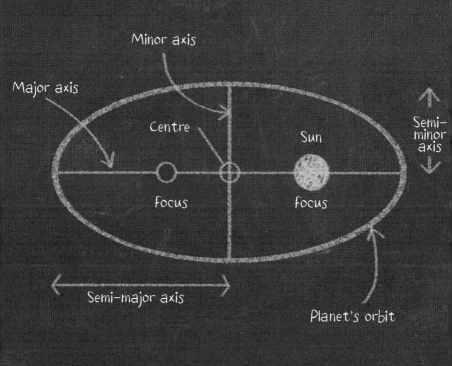

Polynomials

A polynomial is a mathematical expression of the form $a_0 + a_1x + a_2x^2 + ... + a_nx^n$, in which $a_0, a_1, a_2, ...$ are constants. To put it another way, it is a finite series (see page 80) in whole-number powers of x. The highest power in a given polynomial is called its degree. A polynomial of degree 2 goes up to terms in x^2, and is called a *quadratic*. One of degree 3 goes up to terms involving x^3, and is called a *cubic*. Polynomials of degree 1 are called *linear* because their graphs are straight lines. The *zeros of a polynomial* are the solutions to the equation with a polynomial on the left side, and a right-hand side equal to zero.

Polynomials are good local approximations of many functions, and can be used in models for a wide variety of applications, from physics and chemistry to economics and social science. In mathematics, they are important in their own right, and are also used to describe properties of matrices (see page 258) and to create knot invariants (see page 368). Polynomials also play an important part in abstract algebra.

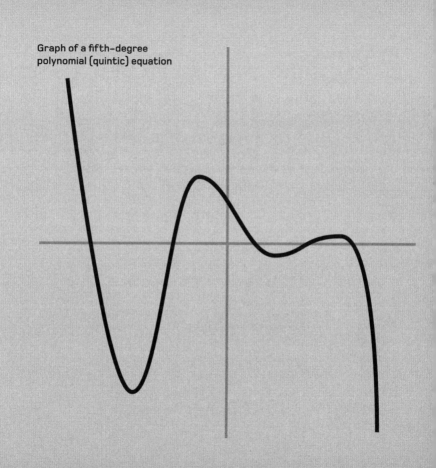

Graph of a fifth-degree
polynomial (quintic) equation

Quadratic equations

A quadratic equation is an equation involving terms up to the squares of a variable, so it is an equation for the zeros of a second-degree polynomial. Geometrically this corresponds to the intersection of a parabola with the x-axis ($y = 0$), and the general form of a quadratic equation is $ax^2 + bx + c = 0$, where a is non-zero.

If $b = 0$, then solving the equation is easy. Rearranging $ax^2 + c = 0$ we find that $ax^2 = -c$, or $x^2 = \frac{-c}{a}$. Our solution is therefore $x = \pm\sqrt{\left(-\frac{c}{a}\right)}$. Note the $\pm$ symbol, indicating that there are positive and negative solutions, both of which can be squared to give a result of $\frac{-c}{a}$. Of course if $\frac{-c}{a}$ is itself negative, we will be unable to find a real-number square root.

A slightly more general argument makes it possible to derive the well-known formula shown opposite. The quantity $b^2 - 4ac$ is called the discriminant of the equation, and its sign deteremines how many real solutions the equation has.

$$x = \frac{-b \pm \sqrt{b^2 - 4ac}}{2a}$$

Cubics, quartics and quintics

Cubics are polynomials where the highest power is 3, so they are third-degree polynomials. Quartics and quintics are fourth- and fifth-degree polynomials, involving a variable raised to the power 4 and 5, respectively. Just as quadratic equations form parabolic curves with a single turning point, so higher-degree polynomials generally define a curve with up to one fewer turning points than their degree. Cubic curves can have two turning points, quartics can have three, and so on.

Finding a generalized solution for these higher-degree equations in terms of elementary functions is much harder than for quadratics. The solution for cubics was discovered in Italy during the 16th century, and these equations turned out to have either one, two or three real-number solutions. A fiendishly clever argument then made the general quartic solvable as well. Quintics eluded all attempts to solve them until the 1820s, when it was proved that there is no general solution for polynomials with degree greater than 4.

Cubic equation

Quartic equation

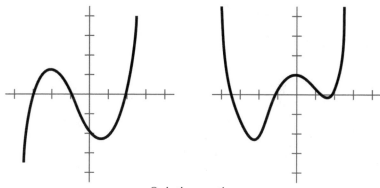

Quintic equation

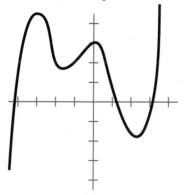

The fundamental theorem of algebra

Fundamental theorems are results that are deemed to have a particular depth and centrality to a field of mathematics. The fundamental theorem of algebra describes the zeros of a general polynomial, and confirms the suspicion from quadratic and cubic equations that the number of real solutions to an nth-degree polynomial equation is bounded by n. It does this by extending our understanding of polynomials beyond those with real-number coefficients, into those with complex-number coefficients (see page 288).

The fundamental theorem gives a factorization of polynomials similar to the prime factorization of numbers (see page 30). It states that

$$a_0 + a_1 x + a_2 x^2 + \dots + a_n x^n$$

can be written as a product of n terms:

$$a_n(x - z_1) \ldots (x - z_n)$$

where $z_1, \ldots, z_n$ are complex numbers, some of which may have zero imaginary part and hence be real numbers. If the coefficients a_i of the polynomial are all real numbers then the complex numbers with non-zero imaginary parts come in complex conjugate pairs (see page 290).

If the polynomial equals zero, then at least one of the terms in brackets must be zero, and vice versa. So, this formula tells us that an nth-degree polynomial has n solutions or roots, though some may be repeated and some may not be real numbers. A repeated root is one that appears more than once, for example $(x - a)^2 = 0$ has one solution, a, but it is repeated twice, once for each bracket.

This result is credited to the great German mathematician Karl Gauss, who published it in 1799. However, there was a hole in Gauss's proof, and it was only rigorously completed in 1920.

Introducing functions

Functions represent relationships between mathematical variables. They take an input, manipulate it in some way, and produce an output. For example, the function $f(x) = x + 2$ takes an input of a number x and produces an output $f(x)$ of two more than x. More sophisticated examples include trigonometric functions, polynomials and power series, but it is hard to do any mathematics without assuming some functional relationship between variables.

A function need not be defined for *all* values of x. It may be specified only for some subset of values, called the *domain* of f. The spread of possible outputs of a function is its *range*. The collection of actual outputs generated by the function acting on a subset of the domain is the *image*.

Despite their importance, remarkably few elementary functions can be easily defined and used. Most others are represented or approximated using these elementary expressions.

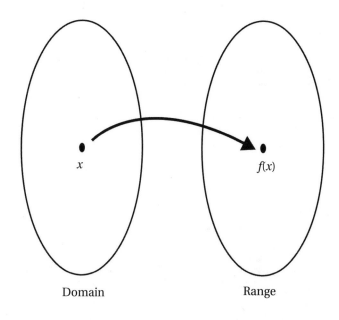

x

$f(x)$

Domain

Range

A function maps any input x in a valid space called the function's domain onto an ouput $f(x)$ in another space called the range.

The exponential function

The exponential function is probably the most important function in mathematics, along with the identity function x. Written $\exp(x)$, it is always positive and tends to zero as x tends to minus infinity, and infinity as x tends to infinity. The graph of $y = \exp(x)$ gets steeper as x grows larger, and the slope of the graph equals the value of the function, that is the height on the y-axis.

The behaviour of phenomena as diverse as radioactive decay, epidemics and compound interest are all described by the exponential function, and it is a building block for many other functions. $\mathrm{Exp}(x)$ is sometimes written as e^x – it is Euler's constant raised to the power x (see page 42). It can also be defined as a power series:

$$1 + x + \frac{1}{2!}x^2 + \frac{1}{3!}x^3 + \frac{1}{4!}x^4 + \ldots$$

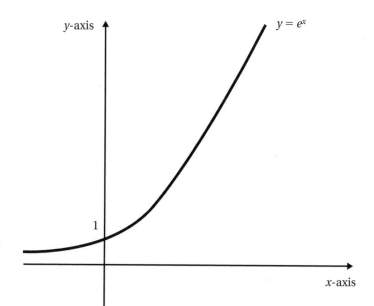

The graph of the exponential function starts off with a shallow slope, but rapidly grows steeper as the value of x increases.

Inverse functions

Inverse functions reverse the action of another function. For example if $f(x) = x + 2$, then the inverse function, known as $f^{-1}(x)$, is $f^{-1}(x) = x - 2$. Inverse functions can be found graphically by reflecting the graph of the original function in the diagonal line $y = x$.

The inverse of the identity function x is x itself, and the inverse function of the exponential function is the natural logarithm (see page 44). The natural logarithm of a given number x, usually written $\ln(x)$, is therefore the power to which e must be raised in order to equal our input x. The natural logarithm also arises as an area, and hence in integration (see page 216): $\ln(n)$, is the area under the curve $y = \frac{1}{x}$, from 1 to n.

Among its many interesting characteristics, the $\ln(x)$ function can be used to describe the approximate number of primes less than x (see page 392).

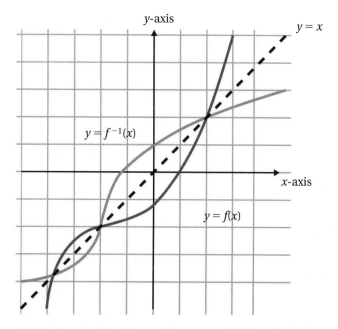

Graph of a sample function and its inverse: note that the graph of the inverse function is identical to that of the original function reflected across the diagonal $y = x$.

Continuous functions

Continuity expresses the idea that the graph of a function can be drawn without lifting your pen from the paper. Conversely to draw a function that is discontinuous, you must lift your pen from the paper. The property of continuity provides a lot of control over the function which makes it possible to make statements about continuous functions as a whole.

If a function is continuous, we can ask questions about how rapidly it varies. Small changes in the variable naturally create only small changes in the output of the function. By choosing a variable sufficiently close to x, we can guarantee that the change in output due to a change in x is as small as we want.

This idea is similar to those used in finding the limits of sequences and series (see page 82), and this is no coincidence. One formal definition of continuity at x is that, for any given sequence of points that converges to x, the sequence obtained by evaluating the function at those points converges to $f(x)$.

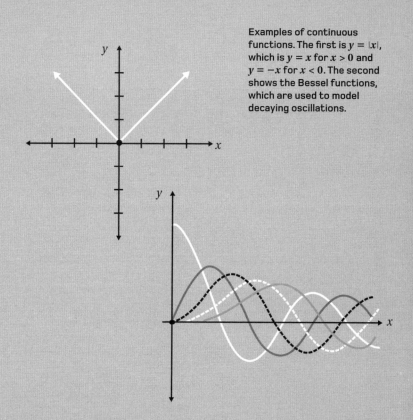

Examples of continuous functions. The first is $y = |x|$, which is $y = x$ for $x > 0$ and $y = -x$ for $x < 0$. The second shows the Bessel functions, which are used to model decaying oscillations.

Trigonometric functions

The elementary trigonometric functions are the sine, cosine and tangent functions, written as $f(x) = \sin x$, $f(x) = \cos x$ and $f(x) = \tan x$. In geometry, the values of $f(x)$ are obtained by a formula involving the angles and sides of a right-angled triangle. However, these functions can be extended using geometric arguments to be defined for all real values of the 'angle'. This heralds the opportunity to see applications for these functions beyond geometry.

When represented as a graph, the sine and cosine functions exhibit a regular pattern, where their shape is repeated every 2π or $360°$. Functions with this repeating pattern are called periodic. This makes them a useful tool in the study of oscillating physical phenomena, such as sound or light waves.

Sine is described as an odd function, with $\sin(-x) = -\sin x$. Cosine, on the other hand, is even, and $\cos(-x) = \cos x$. The output values of both functions always lie between $+1$ and -1.

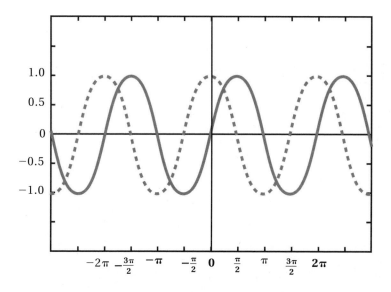

This graph shows how the functions **sin** x (solid line) and **cos** x (dotted line) extend beyond the limits of triangular angles.

The intermediate value theorem

The intermediate value theorem formalizes the idea that a continuous function can be drawn without taking your pen off the page. It states that for any continuous function, given any number between two outputs, there exists an input which has that number as its output, i.e. there are no jumps that miss out some of the possible outputs. For example, if the inputs 10 and 20 give outputs of 20 and 40, the intermediate value theorem implies that for any output of our function between 20 and 40 there must exist an input between 10 and 20 which has that number as an output. Note that while the theorem applies to all continuous functions there are many discontinuous functions that also satisfy it.

The intermediate value theorem is used in many proofs, including the existence of solutions to some equations. It is also an important ingredient of the *ham sandwich theorem*, which states that a piece of ham between two pieces of bread can be simultaneously cut in half with a single slice.

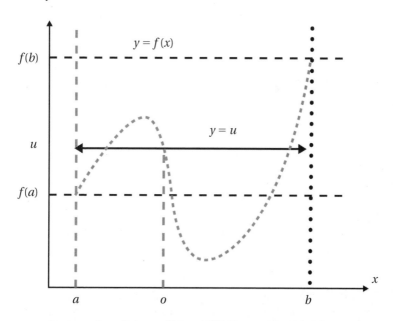

For any value u between $f(a)$ and $f(b)$ there exists at least one value of x between a and b such that $f(x) = u$. For the choice of u in this diagram, there are actually three values of x with this property.

Introducing calculus

Calculus is the branch of mathematics concerned with studying change. Its two fundamental components are *differentiation* (rates of change) and *integration* (sums over changing objects), both of which involve working with infinitely small changes of functions and hence limits. Calculus is a underlying tool of mathematical modelling where rates of change, such as speed, acceleration or diffusion, can be expressed in mathematical form.

The unifying idea behind calculus is that for many functions there is a nice relationship between small changes in outputs and small changes in inputs. Calculus is built out of these relationships. Most classical applied mathematics relies on calculus and functions: phenomena such as waves in fluids, oscillations in mechanics, the motion of the planets, pattern formation on seashells, schooling of fish, chemical changes and forest fires are all described using calculus.

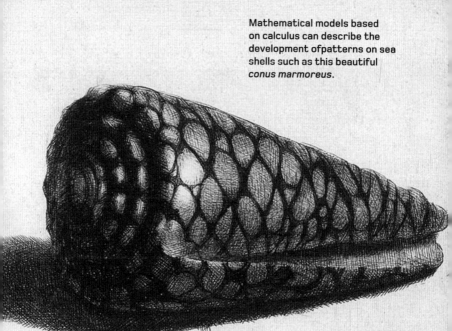

Mathematical models based on calculus can describe the development of patterns on sea shells such as this beautiful *conus marmoreus*.

Rates of change

We can measure the rate of change of a function by using graphs. If the graph of a function is steep, then the output value changes rapidly. If it is shallow, then the output changes more slowly. This has a physical analogy with real hills and valleys, where large gradients mean the altitude changes rapidly as a function of horizontal distance.

For a straight line there is a constant slope or gradient, which is indicated by the quantity m in the line equation $y = mx + c$ (see page 172). For a more general graph we could think about the gradient at a point on the graph as the slope of a tangential line that grazes the graph at that point.

This can be approximated by drawing connecting lines between the point itself and nearby points on the graph, and seeing whether they tend to a limit. When such a slope does exist, it is called the *derivative* of the function at the point, and varies as the point at which we evaluate the slope changes.

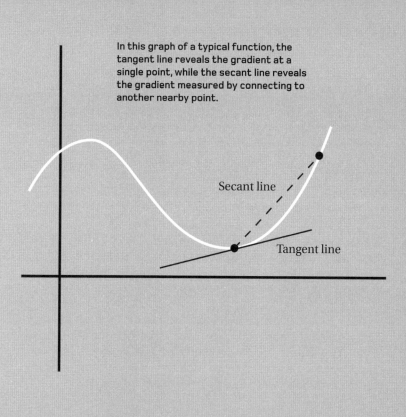

In this graph of a typical function, the tangent line reveals the gradient at a single point, while the secant line reveals the gradient measured by connecting to another nearby point.

Secant line

Tangent line

Differentiation

Differentiation is a key concept of calculus. It is a way of using equations to calculate the slope or gradient of a function, and therefore its rate of change, at a certain point.

The simplest relationship between two variables is a linear one, $f(x) = mx + c$, where m represents the slope. If we fix a value on the x-axis of x_0, then the slope of the function at any point x is related to the amounts of change in the x and the y or $f(x)$ directions. These quantities are represented as $x - x_0$ and $f(x) - f(x_0)$ respectively. Finding the slope at x_0 is a matter of finding the value of m for which $f(x) - f(x_0)$ is approximately equal to $m(x - x_0)$, as x tends towards the value of x_0.

If the limit as x tends to x_0 of the slope m exists, then we say that f is *differentiable* at x_0, and that this limit is the *derivative* of f at x_0. If f is differentiable then the value of m will vary with the value of x_0. In other words, we have created a new function of x, called the derivative of f and written as $\frac{df}{dx}$ or $f'(x)$.

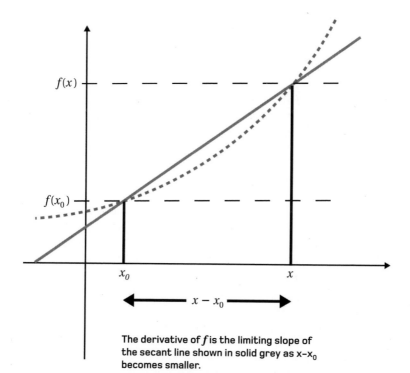

The derivative of *f* is the limiting slope of the secant line shown in solid grey as x–x₀ becomes smaller.

Sensitivity analysis

Sensitivity analysis allows mathematicians and others to measure not just the rate of change, but its significance. For instance, the valuation of pension portfolios involves a balance between current assets and future liabilities. Even if assets and liabilities balance at a particular interest rate, a small future change in that rate may radically change the position. Other examples of such sensitivity include employment figures, climate models and chemical reactions.

The larger the derivative of a function, the faster its rate of change, but a large change to a very large quantity may be less important than a small change to a very small quantity. So a proper assessment requires us to consider both the derivative of the function and its value. One way of doing this uses the *duration* of a function, which is the relative change in its value due to a small change from a current value. This is related to the function's *elasticity*, a term that describes how the slope varies compared with the slope of a linear function.

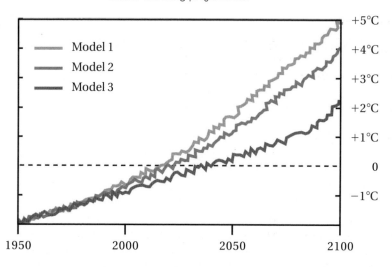

Global warming projections

Model 1
Model 2
Model 3

+5°C
+4°C
+3°C
+2°C
+1°C
0
−1°C

1950   2000   2050   2100

Three graphs of global warming projections for the next century
show the very different results caused by sensitivity due to
behaviour within each model.

Calculating derivatives

The derivative of a function $f(x) = x^n$ is given by the expression:

$$f'(x) = nx^{n-1}$$

where n represents the power to which the original value of x was raised. So the derivative of x^2 is $2x$ and the derivative of x^5 is $5x^4$. Other common examples are given opposite.

If the function $f'(x)$ is itself differentiable, than we can repeat the process and find the second derivative of f:

$$f''(x) = n(n-1)x^{n-2}$$

Continuing in this way, higher and higher derivatives may be calculated. The nth derivative of a function $f(x)$ is denoted as $f^{(n)}(x)$.

f	f'
$\sin x$	$\cos x$
e^x	e^x
$\cos x$	$-\sin x$
x^n	nx^{n-1}

Combining functions

There are two main ways of combining functions to create new functions. The *product* of two functions $f(x)$ and $g(x)$ is obtained by multiplying the values of the functions, forming the function $f(x)g(x)$. For example, the function $x^2\sin x$ is the product of the function $f(x) = x^2$ with the function $g(x) = \sin(x)$.

The *composition* of two functions is obtained by applying them consecutively to get $f(g(x))$. It is sometimes called the function of a function. For the example above, $f(g(x))$ would be $f(\sin x)$ or $(\sin x)^2$. This is different from combining the functions in the opposite order, since $g(f(x))$ would equal $\sin(x^2)$.

The derivatives of products and compositions are found using the product and chain rules, illustrated. Both only hold if the derivatives of the underlying functions exist. The quotient rule, which gives the derivative of one function divided by another, is a direct consequence of the product and chain rules.

PRODUCT RULE

$$\frac{d}{dx} u(x) v(x) = u'(x) v(x) + u(x) v'(x)$$

e.g. $(x \sin x)' = \sin x + x \cos x$

CHAIN RULE

$$\frac{d}{dx} u(v(x)) = v'(x) u'(v(x))$$

e.g. $\left(\sin \left(\frac{1}{3} x^3 - x \right) \right)' = (x^2 - 1) \cos \left(\frac{1}{3} x^3 - x \right)$

QUOTIENT RULE

$$\left(\frac{u(x)}{v(x)} \right)' = \frac{(u'(x) v(x) - u(x) v'(x))}{v(x)^2} .$$

Integration

The process of integration roughly corresponds to finding the area under a graph, but with areas below the axis contributing a negative area. Consider a curve between two points a and b: if we split the area beneath it into thin segments, then the area of each segment is approximately equal to the value of the function at that point, multiplied by the segment width.

Summing these areas together we get an approximation to the total area under the curve. The greater number of ever-narrower segments we use in this process, the more accurate our answer will be. If the limit as the length of the segments goes to zero exists, it is called the *integral* of the function between the limits a and b with $a < b$, denoted

$$\int_a^b f(x)\,dx$$

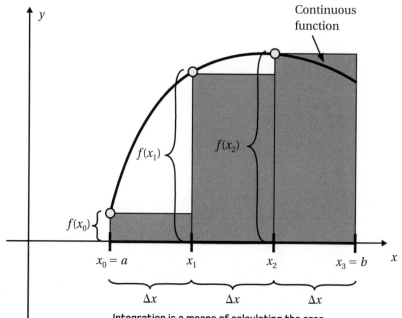

Integration is a means of calculating the area underneath a graph described by $f(x)$. It can be seen as the sum of the areas of a series of columns of width Δx as Δx tends towards zero.

The fundamental theorem of calculus

The fundamental theorem of calculus states that integration is the opposite of differentiation. It uses the idea that the integral of a function f can be thought of as a new function, $F(x)$ say, of the upper limit of the integral, leaving the lower limit unspecified. So $F(x) = \int^x f(u)du$. By convention this is often written as $F(x) = \int f(x)dx$. $F(x)$ is called an *indefinite integral*, and since the lower limit is not specified it is only defined up to a constant, called the *constant of integration*.

Changes in $F(x)$ reflect changes in the area under the curve due to small changes in the upper limit. Since the derivative of a constant is zero, the derivative of the function $F(x)$ does not depend on the constant of integration, and it turns out that it equals the original function $f(x)$. So $F'(x) = f(x)$. This is the fundamental theorem of calculus. A related result is that $\int f'(x)dx = f(x) + c$ where c is a constant of integration. This is a useful way of evaluating many integrals.

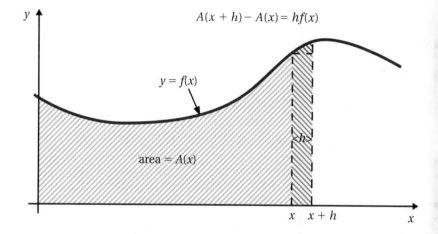

A geometric proof of the fundamental theorem. The area of
the narrow shaded strip can be estimated as $h \times f(x)$ or, if
the function is $A(x)$, it can be computed as $A(x + h) - A(x)$.
Setting these in an equation and dividing both sides by h
reveals that, as h approaches zero, $f(x) = A'(x)$.

Integration and trigonometry

The integrals of some elementary functions of x turn out to be related to trigonometric functions. This demonstrates just how central the trigonometric functions are to mathematics: if they had not been introduced in geometry via the ratios of sides of triangles (see page 132) they would need to have been defined in calculus as the integral of some relatively simple function. An example is:

$$\int \frac{1}{(1 + x^2)}\, dx = \tan^{-1} x + c$$

and another example is given opposite. Here $\tan^{-1}$ is the inverse function of the tangent function, known as arctan. Similarly, $\sin^{-1}$ is the inverse sine function, arcsin. Note that the inverse function is not the same as the reciprocal, $\frac{1}{\tan x}$.

The standard way of deriving these expressions is to use the relationship $\int f'(x)\,dx = f(x) + c$ and the rather easier statement that the derivative of the arctan function, $\tan^{-1} x$, is $\frac{1}{(1 + x^2)}$.

f	f'	$\int f'(x)\, dx$
$\sin x$	$\cos x$	$\sin x + c$
e^x	e^x	$e^x + c$
$-\cos x$	$\sin x$	$-\cos x + c$
$\left(\dfrac{1}{n+1}\right) x^{n+1}$	$x^n \ (n \neq -1)$	$\left(\dfrac{1}{n+1}\right) x^{n+1} + c$
$\ln x$	$\dfrac{1}{x}$	$\ln x + c$
$\sin^{-1} x$	$\dfrac{1}{\sqrt{(1-x^2)}}$	$\sin^{-1} x + c$

Taylor's theorem

Taylor's theorem states that if a function $f(x)$ can be differentiated an infinite number of times then it can be approximated by a power series called a Taylor series. The Taylor series of a function about a point x_0 is a sum of terms involving $(x - x_0)$ raised to successively higher powers of natural numbers.

For a value of x close to 0, the series is:

$$f(x) = f(0) + f'(0)x + \tfrac{1}{2}f''(0)x^2 + \text{...} + \tfrac{1}{n!}f^{(n)}(0)x^n + \text{...}$$

where $f^{(n)}$ is the nth derivative of the function and ! is the factorial operator (see page 104). This special case of a Taylor series is known as a Maclaurin series.

If the series converges (see page 90) for all values of x close to x_0, the function is said to be analytic at x_0. Analytic functions are important in complex analysis (see page 302).

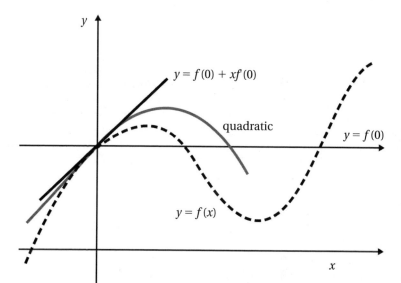

$y = f(0) + xf'(0)$

quadratic

$y = f(0)$

$y = f(x)$

This graph shows successive approximations of a function $f(x)$ by truncated Taylor series about $x = 0$. The quadratic curve (a parabola) uses the first three terms, up to and including the x^2 term.

Interpolation

Interpolation is the art of estimating the output of a function at a specific point, based on the value of that function at other known points. It is important in applications where data is used to build a functional relationship between quantities.

Imagine that we know a function's value $f(x)$ at $n + 1$ points $x_0, x_1, ..., x_n$, with the x_i ordered from smallest to largest. What value should we assign the function at a general point x between x_0 and x_n? This problem arises daily when forecasting weather based on data at a discrete set of sites around the country. One way is to attempt to fit a polynomial (see page 184) through the data points. There are $n + 1$ points and an nth-order polynomial has $n + 1$ coefficients to assign, so there are precisely the right number to match the known values.

The 18th-century French mathematician Joseph-Louis Lagrange found an explicit formula for this form of interpolation, with an associated error similar to that of a truncated Taylor series.

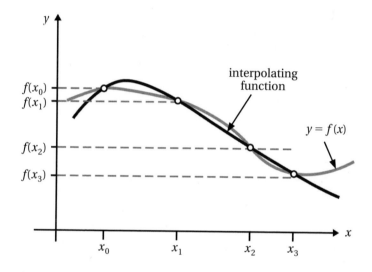

Maxima and minima

The process of finding the maximum or minimum values of a function is called optimization. A maximum of a function $f(x)$ lies at point c if $f(c)$ is greater than or equal to $f(x)$ for all other values of x. Similarly, a minimum lies at d if $f(d)$ is less than or equal to $f(x)$ for all other x. A local maximum or minimum is one where $f(x)$ is only compared for nearby values of x.

At these points, the tangent to the curve is horizontal, so the derivative is zero. This provides an easy way to determine local maxima or minima. At a point c where the derivative is zero, the linear term of the Taylor series (see page 222) disappears and

$$f(x) \approx f(c) + \tfrac{1}{2}f''(c)(x - c)^2 + \text{higher order terms}$$

If $f''(c) \neq 0$ then this is locally like a parabola, with a maximum if the second derivative is negative and a minimum if it is positive. If $f''(c) = 0$ it may instead be a *point of inflection*, where the function flattens out before continuing in the same direction.

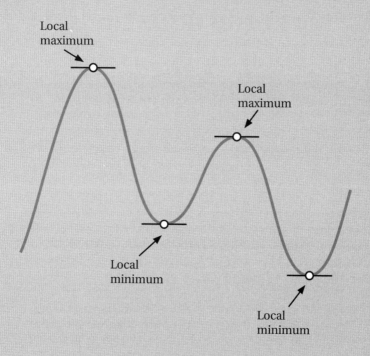

Local maximum

Local maximum

Local minimum

Local minimum

Differential equations

Differential equations express relationships between functions and derivatives. They are used to model many processes in economics, biology, physics and chemistry, where they link the rate of change of a quantity to the quantity itself.

For example, the rate of radioactive decay in a chemical sample is proportional to the number of atoms in the sample, as shown by the equation for radioactive decay: $\frac{dN}{dt} = -aN$, where N is the number of atoms, a is a constant related to the half-life of the element and t is time. This has the solution $N(t) = N(0)e^{-at}$. The expression incorporates a term of the form e^x, showing that the decay is exponential.

Ordinary differential equations are those which involve only one independent variable, such as time in the example above. It is not usually possible to solve them explicitly, so either approximation methods or numerical simulations must be used.

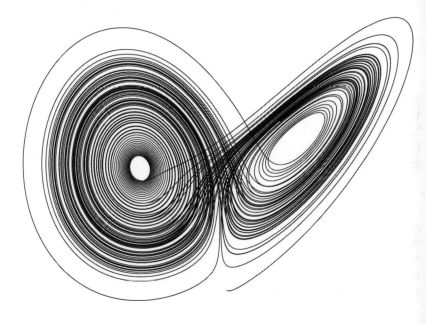

The curve in this picture represents a solution of the Lorenz equations, a differential equation modelling the weather. The curve does not repeat itself and has a fractal structure, indicating the presence of chaos.

Fourier series

Fourier series are functions that are expressed as an infinite sum of sines and cosines. Since the sine and cosine functions consist of repeating patterns, the Fourier series itself is also a repeating, or periodic, function.

For values of x between 0 and 2π we can approximate the function $f(x)$ as:

$$f(x) = a_0 + \sum_{n=1}^{\infty}(a_n \cos nx + b_n \sin nx)$$

where

$$a_n = \frac{1}{\pi}\int_0^{2\pi} f(x)\cos kx \, dx \qquad b_n = \frac{1}{\pi}\int_0^{2\pi} f(x)\sin kx \, dx$$

If the original function is not itself periodic, then the Fourier series provides a representation of the function in the specified interval of values, but not outside it. Instead, it repeats the function, as shown opposite.

Example of a Fourier series:
$f(x) = 1 - x^2$ on $[-\pi, \pi]$

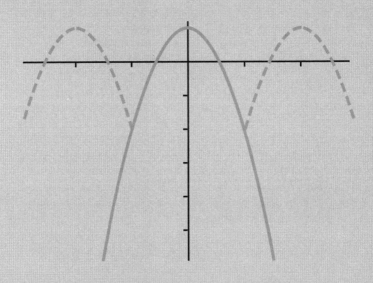

Functions of more than one variable

Functions of more than one variable represent relationships between several different mathematical variables. For example the function $f(x, y) = x^2 + y^2$ is a function in the variables x and y. It takes an input x and an input y and produces an output $f(x, y)$ equal to the sum of their squares.

Equations like this allow us to model functions in three or more dimensions. For example, for Cartesian coordinates (x, y) in the plane, our function becomes a function of these coordinates. We can write this as $f: R^2 \rightarrow R$, to show that the domain of the function is the plane R^2 and the image of the function is the real numbers R. Whereas functions of a single variable can be represented as graphs, three-dimensional functions like this can be represented as surfaces.

These ideas can be extended further, to functions of n real variables, $f: R_n \rightarrow R$, such as $f(x_1, ..., x_n) = x_1^2 + ... + x_n^2$.

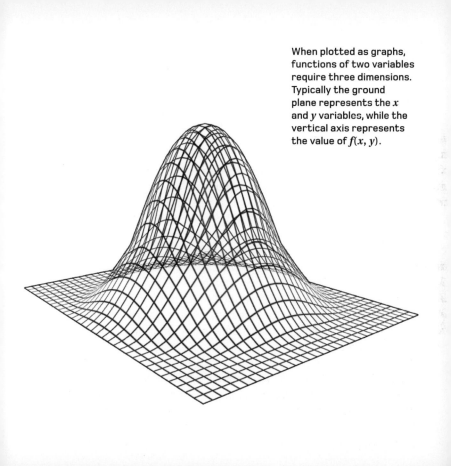

When plotted as graphs, functions of two variables require three dimensions. Typically the ground plane represents the x and y variables, while the vertical axis represents the value of $f(x, y)$.

Partial differentiation

Partial differentiation is the generalization of differentiation to functions of several variables. As with one-dimensional differentiation, the idea is to consider the rate at which a function changes at a specific point. But here there are many different ways of varying the initial point. One choice, in the (x, y)-plane, is to keep y fixed and change x. This defines the partial derivative with respect to x, written $\frac{\partial f}{\partial x}$, which can be calculated exactly as the standard derivative with respect to x, treating y as though it was a constant.

Similarly, the partial derivative with respect to y, $\frac{\partial f}{\partial y}$ is obtained by differentiating with respect to y keeping x fixed. These partial derivatives describe the effect of small changes in two particular directions. The effect of variations in other directions can be obtained from these by using weighted sums of the partial derivatives in x and y, or more generally by using the vector gradient of the function, $\mathbf{grad}(f)$, often represented by the symbol ∇ (see page 252).

$$\frac{\partial f}{\partial x}$$

Integration on a curve

Integrating a function along a curve is the equivalent of integration in one real variable but for functions of more than one variable. In two dimensions, the function $z = f(x, y)$ forms a surface. Imagine a curve in the (x, y)-plane, $z = 0$, and a curtain-like surface connecting this to the surface $z = f(x, y)$ vertically above or below it. The integral of the function along the curve is essentially just the positive or negative area of this curtain, and is sometimes called a line integral.

If y is fixed at a given number, then $f(x, y)$ becomes a function of x and some constant. Thus for fixed y, it is possible to integrate $f(x, y)$ with respect to x, using standard techniques. Equally, if x is fixed, the function can be integrated with respect to y. Geometrically, this corresponds to integrating along straight lines in the (x, y)-plane. There are some technical issues about how and when this can be done in general, but the point is that the idea of integration is easily extended. This is important, for instance, in mechanics, where it is used to calculate work done.

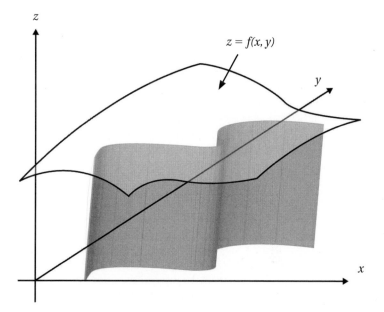

Integration on a surface

Integrating a function on a surface is a higher-dimensional version of integration that creates volumes instead of areas. Imagine a region A in the (x, y)-plane and a function $z = f(x, y)$. By breaking up the area into many very small pieces, the volume under the curve is approximately the value or height of the function at some point, multiplied by the small area. Adding these volumes up gives an approximation of the entire volume under the surface. If this sum tends to a limit as the area of the small pieces tends to zero then this is the surface integral of f on A, denoted:

$$\iint_A f(x, y) \, dx \, dy$$

This is called a double integral, as the area is a product of small changes in x and y. Higher multiple integrals, giving generalized integration for functions of more variables, can also be defined.

The double integral of the function over the rectangular area gives the volume of the shaded area.

Green's theorem

Green's theorem links a double integral over a surface A with a line integral around the surface's boundary γ. It states that:

$$\iint_A \left(\frac{\partial f}{\partial x} - \frac{\partial f}{\partial y} \right) dxdy = \int_\gamma fds$$

where ds denotes the one-dimensional small change along the path of γ.

Equations like this hint at a very abstract connection between generalized integrals and partial derivatives. Vector-valued functions provide several more key examples (see page 252). Given the fundamental theorem of calculus, such connections should not be a complete surprise. The interesting point here is that the connections between integrals over surfaces and curves generalize to statements about integrals over n- and $(n-1)$-dimensional surfaces.

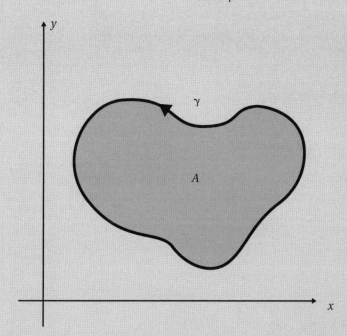

A region A and its boundary curve γ.

Introducing vectors

Vectors are used to represent mathematical or physical quantities which have magnitude, or length, and direction. For example, the wind has a given speed and direction, and like winds on a weather map, vectors are often represented by an arrow, with the arrowhead and alignment of the arrow defining the direction, and the length denoting the vector's size.

Once you understand how vectors combine, and the intuitive meaning of those combinations, geometric calculations that would be very complicated without vectors become routine. Thus vectors provide another set of techniques for tackling geometric problems, and having different ways to approach the same mathematical problems can lead to new insights. Because the algebraic structure of vectors is mimicked by many other mathematical objects, they are extremely useful. Collections of vectors known as vector spaces can be applied in many areas of mathematics, and have wide-ranging applications in science and engineering.

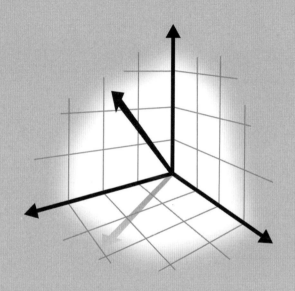

Adding and
subtracting vectors

Adding two vectors is as simple as placing their arrows nose to tail, and drawing a new arrow from the start point to the end point. This new vector is known as the *resultant vector*.

Vectors can also be described by Cartesian coordinates, where the point (x, y) gives the position of the end point with respect to an arbitrary origin. Just like following a treasure map, if we move x steps in the x-direction, then y steps in the y-direction, we will reach our goal. The sum of two vectors $(1, 0)$ and $(0, 1)$ can therefore be calculated by adding the coordinate components independently, giving $(1, 1)$. Subtraction works in the same way: the resultant of $(3, 2)$ minus $(1, 1)$ is $(2, 1)$.

Because each coordinate of a vector represents one side of a right-angled triangle, its magnitude or *modulus* can be obtained by Pythagoras' theorem (see page 130). The modulus of a vector $(1, 1)$ is equal to the hypotenuse of a triangle with sides of length 1. By Pythagoras theorem, this is $\sqrt{(1^2 + 1^2)}$, or $\sqrt{2}$.

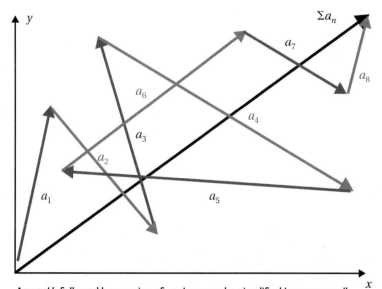

Any path followed by a series of vectors can be simplified to one overall vector with a single direction and magnitude, represented by Σ, the Greek letter sigma.

Scalar product

The *scalar* or *dot product* is an operation that combines two vectors to create a *scalar*, a number without a given direction. Written as $\boldsymbol{a} \cdot \boldsymbol{b}$, it is the product of their lengths multiplied by the cosine of the angle between them. If the vectors are represented in component form, then the scalar product is the sum of the products of each pair of components. The scalar product of $(1, 2)$ and $(1, 3)$ is therefore $(1 \times 1) + (2 \times 3) = 7$.

If two vectors are perpendicular, then the cosine of the angle between them is zero. Hence the scalar product of two perpendicular vectors is also zero. If either vector is a *unit vector*, with magnitude or modulus 1, then the scalar product is simply the component of the other vector in the direction of the unit vector: the scalar product of $(2, 3)$ and $(0, 1)$ is 3.

This concept is important in physics, where properties such as magnetic flux are given by the scalar products of vectors that represent the magnetic field and the area being considered.

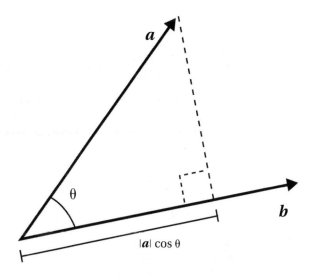

$|a| \cos \theta$ is the projection of a onto the direction of b, so the scalar product, $|a||b|\cos \theta$, is the product of the modulus of the vector b and the projection of a onto b (or vice versa).

Cross product

The vector or *cross product*, written $a \times b$, is a method for multiplying two vectors in three-dimensional space, producing a vector that is perpendicular to both initial vectors. In physics, it can be used to calculate the torque of a force. The magnitude or modulus of a vector product of two vectors is the product of their lengths times the sine of the angle between them. This is also equal to the area of a parallelogram whose adjacent sides are given by the two vectors.

The direction of the resulting vector is given by a convention called the right-hand rule, shown opposite. If the first finger of the right hand represents vector a, and the second finger represents vector b, the direction of the vector product is indicated by the thumb. Using the right-hand rule to identify the direction of both $a \times b$ and $b \times a$, we find the thumb points in opposite directions. So the order in which the vectors are written matters: unlike the normal multiplication of numbers, the vector product is a *non-commutative* property.

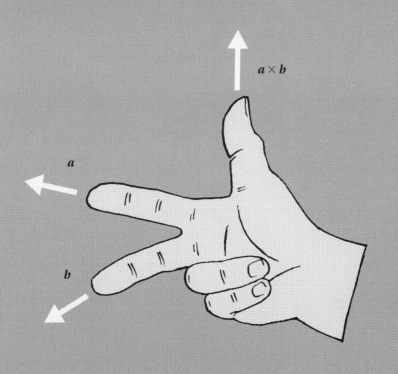

Vector geometry

Vector geometry describes the use of vectors to solve geometrical problems. Many ideas in geometry are greatly simplified by posing them in vector form, especially when working in three or more dimensions. For example, if the position of a point in three dimensions is represented by a vector $r = (x, y, z)$, called the position vector, then a two-dimensional plane through a point with position vector r_0 is given by the solutions of $a \cdot (r - r_0) = 0$, where a is a vector perpendicular to the plane.

If we write out the coordinate equations of three planes using this formula, the condition for their intersection is given by three simultaneous equations (see page 168). The advantage of seeing the problem in this way is that it becomes obvious from the geometry that three simultaneous linear equations either have a unique solution, the *typical* case (infinitely many solutions, where all planes would be the same) or no solutions, where at least two planes are parallel and not equal.

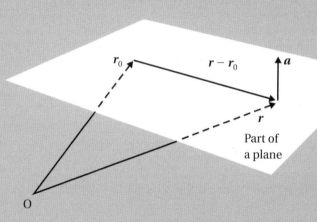

Vector functions

Vectors whose components are functions, describing a relationship between two or more variables, are *vector functions*. To study these relationships, the components can be differentiated or integrated like real functions.

Differentiation itself can be expressed by a vector operator. For example, if $f(x, y)$ is a real function in a plane, the gradient of f is given by a vector function $\left(\frac{\partial f}{\partial x}, \frac{\partial f}{\partial y}\right)$, written as ∇f. This vector's direction and magnitude give the direction of the greatest rate of increase in f, and the rate of that increase.

The operator ∇, known as del, has many beautiful properties. Two that relate integrals are shown opposite. One example is that the flux out of the boundary of a surface is equal to the divergence of the vector function within that surface. This explains what happens when air is pumped into a tyre: since the flow or flux of air *out* of the tyre is negative the expansion of air within the tyre is also negative, in other words, it is compressed.

Divergence theorem:

$$\iint_{\delta V} f.dS = \iiint_V (\nabla.f)dV$$

Stokes' theorem:

$$\int_{\delta A} f.dl = \iint_A (\nabla \times f).dS$$

Dimensions and linear independence

The *dimension* of an object or space is a measure of its size. For standard Euclidian space it is the number of coordinates needed to specify the points within that space. For instance, a circle is one-dimensional, a disc is two-dimensional and a sphere is three-dimensional. Intuitively we understand that there are two or three directions that can be explored: up, down and sideways. This is expressed mathematically using the idea of *independence*.

A set of vectors is *linearly independent* if none of the vectors can be written as the sum of multiples of the others. Any set of n linearly independent vectors is said to be a *basis* for n-dimensional space, and any vector in the space can be written as a linear combination of basis vectors. In three dimensions, the standard Cartesian basis is the set of coordinate vectors $(1, 0, 0)$, $(0, 1, 0)$ and $(0, 0, 1)$, which have the additional property that they are perpendicular to each other. But any three linearly independent vectors is an acceptable basis for three dimensions.

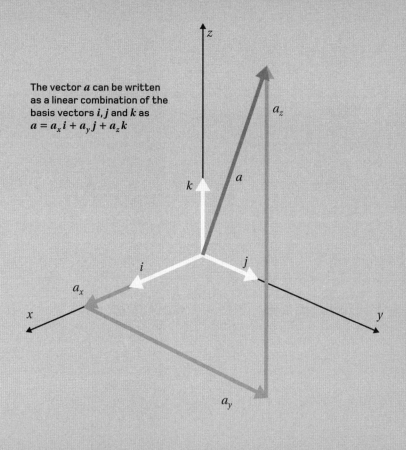

The vector a can be written as a linear combination of the basis vectors i, j and k as
$a = a_x i + a_y j + a_z k$

Linear transformations

A linear transformation is a function that transforms one vector into another vector, while also respecting the rules of linear combination. For example, a transformation applied to the sum of vectors must have the same result as if each vector was transformed, then added together. More generally, if a and b are scalars and u and v are vectors, then the linear transformation L must conform to $L(au + bv) = a(Lu) + b(Lv)$. So if we know the value of the linear transformation on a set of basis vectors, we will also know the value of the transformation everywhere within the basis-defined space.

Linear transformations have a geometric interpretation and include translations, rotations and shears. Thus the language of linear transformations provides a way to describe simple geometric operations. They also arise naturally in calculus: in fact, derivatives (see page 208) are nothing more than linear transformations on functions, and the study of linear transformations unifies aspects of geometry and calculus.

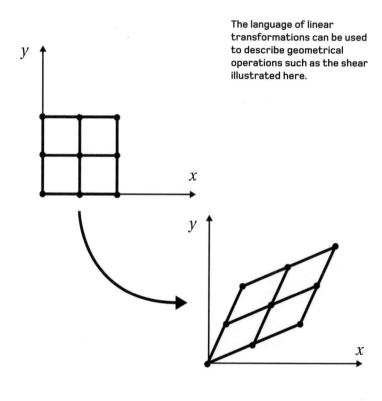

The language of linear transformations can be used to describe geometrical operations such as the shear illustrated here.

Introducing matrices

Matrices are collections or arrays of numbers organized in fixed numbers of rows and columns. They are written within brackets as, for example: $\begin{pmatrix} 1 & 3 \\ 0 & 2 \end{pmatrix}$ or $\begin{pmatrix} a & b & a \\ c & a & c \end{pmatrix}$.

Matrices can be used in many contexts, but are particularly helpful when calculating the effect of linear transformations. Given a coordinate point (x, y), a general linear transformation takes, or maps, this to a new point $(ax + by, cx + dy)$, a process known as matrix multiplication. We can represent this as Mr where r is the position vector (x, y) and M is a matrix $\begin{pmatrix} a & b \\ c & d \end{pmatrix}$ representing the action of the linear transformation. This 2×2 matrix definition can easily be extended to $n \times n$ for work in higher dimensions.

The *identity matrix*, I, has 1s in the diagonal positions and zeros elsewhere. Ir therefore equals r for any vector r.

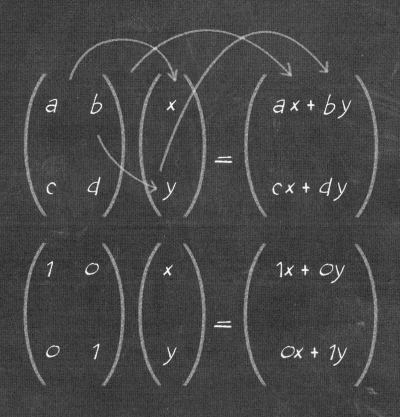

Solving matrix equations

Matrix equations are mathematical equations where an entire matrix array is represented by a single variable. These simplified expressions are used in a wide variety of contexts, including linear transformations.

If $M\boldsymbol{r}$ describes the effect of a linear transformation on a vector $\boldsymbol{r}$, then the vectors that map to a given vector $\boldsymbol{b}$ under the transformation are given by the solutions to the matrix equation $M\boldsymbol{r} = \boldsymbol{b}$. To solve this we need to use the inverse matrix of M, if it exists.

The inverse M^{-1} is the matrix that, when multiplied by M, gives the identity matrix I. Applying M^{-1} to the equation $M\boldsymbol{r} = \boldsymbol{b}$, we can say that $M^{-1}M\boldsymbol{r} = M^{-1}\boldsymbol{b}$. Since $M^{-1}M = I$, then $I\boldsymbol{r} = M^{-1}\boldsymbol{b}$. And since the identity matrix leaves a vector unchanged, $\boldsymbol{r}$ is therefore equal to $M^{-1}\boldsymbol{b}$.

Of course, this is no help unless we know M^{-1}, but at least in the 2×2 case this is easy to calculate. For the general matrix $\begin{pmatrix} a & b \\ c & d \end{pmatrix}$, the inverse is $\frac{1}{ad - bc} \begin{pmatrix} d & -b \\ -c & a \end{pmatrix}$, provided that $ad - bc$ is not equal to zero.

If we think about the coordinate representation of the matrix equation $Mr = b$ it is precisely a set of simultaneous linear equations. Thus we come full circle: the problem of finding the intersection of three planes (see page 174) is equivalent to solving three simultaneous linear equations (see page 168) via the vector representation of planes (see page 250), and is also equivalent to solving matrix equations.

Our matrix inverse allows us to see that in two dimensions, where planes are the same things as lines, there is a unique solution provided that $ad - bc$ does not equal zero. If it does, then there are either no solutions or infinitely many solutions. The quantity $ad - bc$ is called the *determinant* of the matrix. In higher dimensions the expression is more complicated but there are standard ways of calculating it.

Null spaces

Also known as the *kernel* of a matrix, the null space is the set of all vectors that are mapped to the zero vector by the action of the linear transformation. For a matrix M, where $M\boldsymbol{r}$ describes the effect of a linear transformation on a vector $\boldsymbol{r}$, the null space N is the set of points for which $M\boldsymbol{r} = 0$. The dimension of this null space is known as its *nullity*.

To explore the size or dimension of the transformed vectors, consider the *image space* $\text{Im}(M)$: this is the set of points $\boldsymbol{b}$ for which $M\boldsymbol{r} = \boldsymbol{b}$ for some value of $\boldsymbol{r}$. Then the rank of M is the dimension of its image space. Moreover, if $M\boldsymbol{r} = \boldsymbol{b}$ has one solution for a given $\boldsymbol{b}$, then it has a *space of solutions* equal to the dimension of N. This is because adding any vector in N to the known solution is also a solution. So if $\boldsymbol{b}$ is in the image of M there is a solution, and the multiplicity of solutions is described by the dimension of N.

Since the effect of a linear transformation can be deduced from its action on a set of basis elements (see page 256), it should not be a surprise that the size or dimension of the set of points in the image of the transformation, $\mathrm{Im}(M)$, equals the number of linearly independent vectors in the transformed basis elements.

If this number is k, and we are working in n dimensions, then there are $n - k$ linearly independent vectors which map to the zero vector. In other words, the dimension of the image of the transformation (its *rank*) plus the dimension of its null space (its *nullity*) equals the dimension of the vector space we are working in.

This might not seem like a big deal, but it is the sort of *decomposition theorem* that mathematicians love, and it has important consequences. Since many problems, such as linear differential equations, can be expressed in this language, the very precise description of solution spaces from this result is used across several areas of mathematics.

Eigenvalues and eigenvectors

Eigenvalues and eigenvectors are special sets of scalars and vectors associated with a given matrix. Their names derive from the German *eigen*, meaning 'peculiar to' or 'characteristic'. For a square matrix M, with eigenvalue λ and corresponding eigenvector r, then Mr equals λr. In physical terms this means that eigenvectors are those directions that remain unchanged by the action of the matrix M, and λ describes how distances are changed in that direction with negative eigenvalues indicating a reversal of direction.

If we try to solve the equation $Mr = \lambda r$, it is the eigenvalues (λ) that are the easiest to obtain. By rewriting the definition as $(M - \lambda I)r = 0$, we can see that solutions exist only if $(M - \lambda I)$ has a non-trivial null space. This means that the determinant of $(M - \lambda I)$ must be zero. The determinant of such an $n \times n$ matrix turns out to be a polynomial (see page 184) of degree n in λ. Eigenvalue problems are common since they provide a great deal of information about linear transformations.

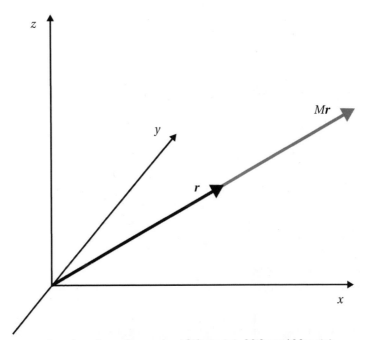

A vector *r* is an eigenvector of the matrix M if *r* and M*r* point
in the same direction (or precisely opposite directions).

Introducing
abstract algebra

Abstract algebra is the study of structure imposed by different rules for combining elements of a set. These rules mimic different aspects of the familiar operations of addition and multiplication for ordinary numbers, and the structures created include groups, fields, rings and vector spaces.

For instance, a vector space is an abstract structure that includes a set of vectors plus their associated rules. These rules describe how combinations of objects within the structure behave, and can be codified as a short list of properties. In vector spaces, the rules describe vector addition (see page 244) and scalar multiplication (see page 246).

This move away from explicit applications in real space towards a more abstract set of properties is typical of the way that mathematicians develop ideas. Despite their abstraction and restriction, these amazing structures have far-reaching implications in areas from molecular structure to topology.

Group theory plays an important role in understanding crystalline structures, since symmetry groups can be used to model the behaviour and possible arrangements of atoms in a crystal lattice.

Groups

A group is a set of elements together with a *binary operation*, which could be thought of as multiplication or addition, but is not named in the general definition.

For any set G, operation $\cdot$ and three elements a, b and c, four basic properties or axioms must be satisfied:

1. *Closure: if a and b are in G, then so is $a \cdot b$.*
2. *Associativity: $a \cdot (b \cdot c) = (a \cdot b) \cdot c$.*
3. *Identity: there exists an element e in G such that $e \cdot a = a$ for all a in G.*
4. *Inverse: for all a in G there exists a^{-1} in G, such that $a \cdot a^{-1} = e$, where a^{-1} is known as the inverse element of a.*

For instance, the set of integers and the operation of addition form a group with $e = 0$, since it is the only number that can be added to an element without changing it. Groups can also be used to represent physical properties, such as the symmetries of regular polygons, crystalline structures or snowflakes.

Symmetry groups

A symmetry group represents the different ways that an object can be transformed so the end result is indistinguishable from the start. It also includes the operation of *composition*, applying one transformation to the result of another, as do all group manipulations.

Consider an equilateral triangle. If we rotate it clockwise by $120°$, or reflect it in a line through a vertex and the centre, the result appears unchanged. If we call the rotation a and the reflection b, then we can use multiplication to indicate compositions of the two.

So, a^2b means we rotate the triangle by $120°$ twice, then reflect it in the line. In fact there are six different combinations of a and b that produce independent transformations of the triangle: e, a, a^2, b, ab and a^2b, where e is the identity, which does nothing to the triangle. Every other combination is equivalent to one of these: a^3 or b^2 are the same as doing nothing, or e.

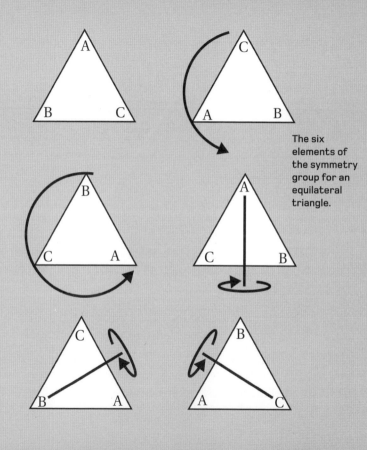

The six elements of the symmetry group for an equilateral triangle.

Subgroups and quotient groups

A subgroup is a subset of a group that satisfies the group axioms (see page 268). Since the identity element $\{e\}$ is itself a group, there is always at least one subgroup.

The group of symmetries of the equilateral triangle (see page 270), is $\{e, a, a^2, b, ab, a^2b\}$ where a is a rotation of $120°$ around the centre and b is a reflection in a symmetry line through the centre. This group has two obvious non-trivial subgroups, the rotations $\{e, a, a^2\}$, and the reflections $\{e, b\}$. Both of these are examples of cyclic groups, where all elements are compositions of a single element.

If H is a subgroup of G, and ghg^{-1} lies within H for all h in H and g in G, then H is called a normal subgroup. Normal subgroups allow us to construct new groups from old ones.

A quotient group is a group constructed from the elements of a group and one of its normal subgroups.

If H is a normal subgroup of the group G then, for any two elements a and b in G, either $aH = bH$, where xH is the set of all points of the form xh for some h in H, or the two sets have no elements in common. This means that we can think of these sets as elements of a new set, and with the natural combination law $(aH)(bH) = abH$, this turns out to be a new group, called the quotient group and denoted G/H.

The quotient group and the normal subgroup that defines it effectively act as a factorization of the group G into smaller groups, which helps us to understand the original group. These smaller groups act as building blocks for the group in the same way that the structure of numbers is described by prime factorization.

For groups, the role of prime numbers is played by the *simple groups*, those that have no non-trivial normal subgroups other than themselves.

Simple groups

Simple groups are groups that do not have non-trivial quotient groups. Their only normal subgroups are either the identity or the original group itself. This is almost exactly analogous to the prime numbers, where a prime has only one and itself as factors.

Just like primes, there are infinitely many simple groups. Unlike primes, however, simple groups can be neatly classified. The classification in 2004 of all the finite simple groups is among the greatest mathematical achievements of the last fifty years.

Simple groups include the cyclic groups of prime order and the family of alternating groups, which arise naturally in the study of finite sets. There are 16 other families of simple groups, called the Lie-type groups, and then 26 exceptions, isolated special cases called sporadic groups. Of these, 20 are related to the largest of the exceptions, the Monster group. The remaining six are known as the pariahs.

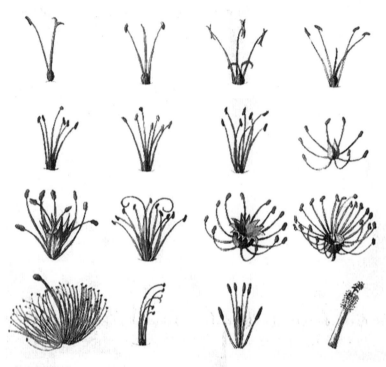

Swedish botanist Carolus Linnaeus' attempts to sort plants by the shape of their reproductive parts is comparable to classifying mathematical groups.

The Monster group

The Monster group is the largest sporadic simple group and is important in the classification of finite groups. Its only normal subgroups are the trivial group and the Monster itself.

Originally conjectured in the 1970s, the Monster was finally hunted down by Robert Griess in 1981, and described in a 1982 paper titled *The Friendly Monster*. It contains **808017424794512 87588645990496171075700575436800000000** (roughly 8×10^{53}) elements. Written in matrix form, the monster array requires $196,883 \times 196,883$ components.

The size and complexity of groups like this mean it has taken time to be sure that all the possible sporadic groups are accounted for. Although the earliest were discovered in the late 19th century, the complete description of all sporadic groups was only completed in the early 21st century.

8080174247
9451287 5886
4599049617
1075700057
5436800000
00000

Lie groups

Lie groups are important families of groups whose elements depend on continuous variables, unlike the discrete structures of the Monster group and the symmetry groups of polygons. For example, if we consider the symmetry of a circle we find that a rotation about the centre by any angle maps the circle onto itself. The symmetry group of the circle can therefore not be classified in the same way as that of a body like an equilaterial triangle, with its six discrete elements (see page 270). The circle's symmetry group, which is a Lie group, is said instead to have *continuous parametrization*.

Unsurprisingly, the theory of continuous groups is more complicated than that of discrete groups, although the Lie groups are the best understood of them. They may be described only through the nature of their parameters, but they inherit more than just their continuous structure. They can also be viewed as smooth, or differentiable, manifolds, which are specific types of topological spaces (see page 336).

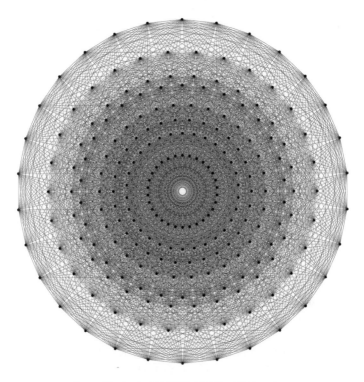

A graphical representation of the E8 Lie group.

Rings

A ring is an abstract mathematical structure that includes a set of elements together with *two* binary operations. This is in contrast to a group, which contains a set of elements and a single binary operation. In the theory of rings, the operations are usually called addition, $+$, and multiplication, $\times$, and as with groups, when either of the operations is performed on two elements of the set, the result should be another element within the set, and hence within the ring.

Unlike groups, where there is no assumption that the operation within the group is commutative (see page 24), the additive operation of a group must be commutative. In other words, for any elements a and b, $a + b$ must equal $b + a$. There must also be an additive identity and inverse, so that the elements of the ring form a group under addition. The multiplicative operation must be associative (see page 24).

Finally, two laws which determine how combinations of the additive and multiplicative operations behave also need to hold. These laws make multiplication distributive over addition:

$$a \times (b + c) = (a \times b) + (a \times c)$$

$$\text{and } (a + b) \times c = (a \times c) + (b \times c).$$

The integers, the rational numbers and the reals are all rings. However, a general ring has properties that are unlike these examples. For instance, if $a \neq 0$, where 0 is the *additive identity*, the element which, when combined with any other element using the addition operation, leaves that element unchanged, and if $a \times b = 0$, then we cannot necessarily conclude that $b = 0$, even though this would be obvious for the rationals, integers or reals. For similar reasons, if $a \times b = a \times c$ then b and c are not necessarily equal either.

Despite these restrictions rings arise naturally in a number of areas of mathematics, particularly those associated with group theory. To allow features such as multiplicative cancellation, further restrictions need to be placed on the algebraic structure, leading to fields (see page 282).

Fields

A field is an algebraic structure that includes a set and two binary operations. As with a ring, these operations are known as addition and multiplication, and likewise the set together with the additive operation forms a commutative group. However, multiplication is also commutative within a field, so for any elements a and b, $a \times b = b \times a$ and the set, except the additive identity element, forms a commutative group with the multiplicative operation. The distributive laws of a ring also hold (see page 281).

This means that division is possible in a field for all elements except the additive identity, and means that, unlike in rings, if $a \times b = a \times c$ and $a \neq 0$, then $b = c$. Thus a field has more of the features that standard numbers have under addition and multiplication than a ring. The integers, rational numbers and reals are all fields, as well as being rings. Another example is the set of numbers of the form $a + b\sqrt{2}$, where a and b are rational.

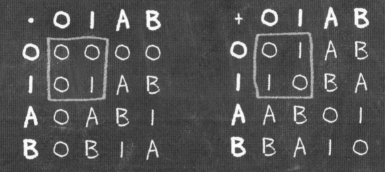

These two tables represent the addition and multiplication operations on a simple field of four elements, I, O, A and B. Here, I is the field's multiplicative identity, and O is its additive identity.

Galois theory

Galois theory was developed by the French mathematician Evariste Galois, who died after a duel aged 20. It connects group theory to the solutions of polynomials (see page 184).

The general solutions to quadratic, cubic and quartic equations were known by the late 16th century, but no such solution had been found for higher-order polynomials. Although solutions to polynomials appear rooted in algebraic manipulation, Galois showed that group theory can reveal whether a polynomial has a closed form solution, involving simple algebraic operations.

Galois looked at the way that equations with given solutions could be transformed between each other, and found that the existence of closed form solutions is related to whether or not an associated group is commutative. Only the first four of the solvable groups he constructed are commutative, indicating that only polynomials up to and including degree 4 can be solved generally in terms of simple algebraic functions.

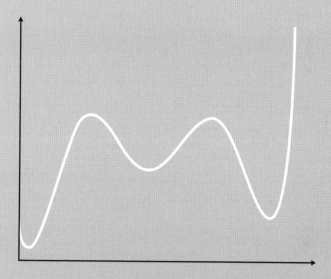

Galois theory proves that finding a generalized solution for sextic equations (polynomials of degree 6), such as the one represented here, is impossible.

Monstrous moonshine

The monstrous moonshine conjectures reveal links between two different areas of mathematics. They were proposed by British mathematicians John Conway and Simon Norton after John Kay mentioned a bizarre coincidence during a 1978 seminar. Kay noticed that a coefficient in the expansion of a function defined in number theory by Felix Klein was 196,884, just one digit from 196,883, the size of the Monster group in matrix form.

The answer to why these two areas – representations of the Monster group on the one hand and algebraic number theory on the other – are so intimately connected uses ideas from yet another area of mathematics, the vertex operator algebras of theoretical physics. In a piece of work that earned him a Fields Medal, the highest award in mathematics, Richard Borcherds showed that the conformal field theory of theoretical physics provides an explanation for this deep connection. However, many details of this relationship between quantum theories, algebra, topology and number theory are still not understood.

Complex numbers

Complex numbers are an extension of the real numbers that make it possible to make sense of the square roots of negative numbers. Any complex number z can be written as $a + ib$, where a and b are real and i is the square root of -1, so $i^2 = -1$: a is the real part of z and b is the imaginary part.

If we think of (a, b) as Cartesian coordinates, we can explore the geometry of complex numbers as in the figure opposite. This is called the Argand diagram. As a point in the plane, any complex number z therefore has a distance from the origin, called the modulus of z and denoted $|z|$. By Pythagoras' theorem $|z|$ can be calculated from its two components, using $|z|^2 = a^2 + b^2$.

Any complex number also has an angle, relative to the x-axis, called the argument of z. A complex number can therefore be specified in terms of its modulus $|z|$ and the angle θ, as $z = |z| (\cos \theta + i \sin \theta)$.

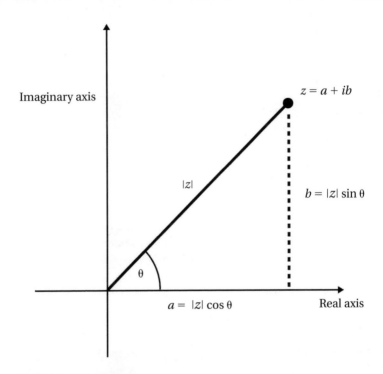

The Argand diagram

Geometry of complex numbers

The geometric interpretation of complex numbers using the Argand diagram provides a simple interpretation of two other features of complex numbers: the complex conjugate and the triangle inequality.

The *complex conjugate* of $z = a + ib$, known as z^* or $\bar{z}$, is $a - ib$, the image of z reflected in the real (x) axis. A simple calculation shows that $|z|^2 = zz^*$, and also that the real and imaginary parts of z can be written in terms of the sum and difference of the number and its complex conjugate as $\frac{(z + z^*)}{2}$ and $\frac{(z - z^*)}{2i}$.

The triangle inequality is the mathematical formulation of the statement that *the long side of a triangle must be shorter than the sum of its other two sides*. The sum of two complex numbers is geometrically the same as the sum of two vectors (see page 244) where the components of a complex number are its real and imaginary parts. Thus, for complex numbers z, w and $z + w$, $|z + w| \leq |z| + |w|$, which is the triangle inequality.

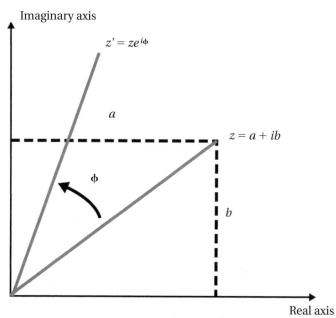

Another consequence of the geometric view of complex numbers is that, since $z = |z|e^{i\phi}$, multiplication by a complex number e^i is, using the rules of exponents, $ze^{i\phi} = |z|e^{i(\theta + \phi)}$, which is just a rotation by an angle θ.

Möbius transformations

Möbius transformations are functions of the complex plane which map circles and straight lines to circles and straight lines. They take the form $f(z) = \frac{(az + b)}{(cz + d)}$ where $ad - bc \neq 0$, and where a, b, c and d are complex numbers and z is a complex variable.

The compositions of these transformations form a group (see page 268), with the group operation being equivalent to matrix multiplication on the 2×2 matrix, with entries a, b, c and d. Crucially, they also preserve angles.

Möbius transformations are used in physics, for example, to change two-dimensional fluid models into simpler scenarios, where problems can be more easily solved, and then back again.

Some features of the group of 2×2 complex matrices can also be visualized using Möbius transformations, creating beautiful graphs such as the one shown opposite.

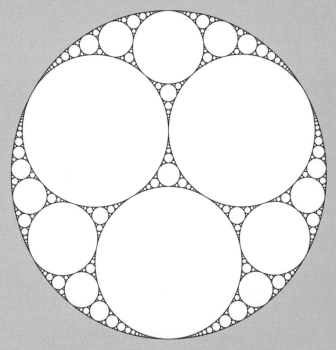

The Apollonian gasket is one of many remarkable patterns
that can be generated using Møbius transformations.

Complex power series

Complex power series, or complex Taylor series, include infinite series of the form $a_0 + a_1z + a_2z^2 + a_3z^3 + ...$, where the coefficients a_k are all complex numbers. More generally, the powers of z can be replaced by powers of $(z - z_0)$ for some fixed complex number z_0.

As with real power series (see page 106), the issue of convergence is central to the theory of power series. One way of establishing convergence is comparing the sum of the moduli of each term $|a_0| + |a_1z| + |a_2z^2| + |a_3z^3| + ...$ with the geometric series $1 + r + r^2 + r^3...$ (see page 100).

If the power series converges for all values of z then the function created by the series is *entire*. Entire functions include the complex polynomials and complex exponential. If the power series converges for values of z close to z_0, then the *radius of convergence* of the series is the largest r such that the series converges for all z within a circle of radius r centred on z_0.

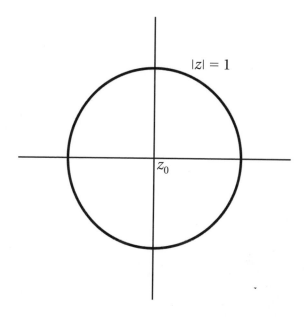

Modulus of a complex power series, showing divergence at some
point and the radius of convergence about a given point z_0.

Complex exponentials

Complex exponentials arise if we take the definition of an exponential (see page 194) and apply it to a complex number, $z = x + iy$. Since the exponential of z, $e^{x + iy}$, can be expressed as $e^x e^{iy}$, where e^x is the standard real exponential, anything new in this quantity is going to come from the imaginary bit, e^{iy}, known as the *complex exponential*.

In fact, representing e^{iy} as a power series (see page 106) and separating out the real and imaginary terms leads to the conclusion that

$$e^{iy} = \cos y + i \sin y.$$

So trigonometric functions are not really geometric in origin – they are actually complex exponentials! This remarkable discovery has important practical uses: it allows engineers to use complex numbers for modelling alternating current, and physicists to use complex wave functions to describe the probability of events in quantum mechanics.

Mathematically it is perhaps more natural to start with the exponential function and complex numbers and deduce the geometric interpretation. Note that, using the corresponding formula for e^{-iy}, both cosine and sine functions can be written as sums or differences of the exponentials themselves.

The relationship between complex exponentials and the sine and cosine functions gives rise to what many consider the most beautiful equation in mathematics. This is *Euler's identity*, which links the five most important numbers in analysis: 0, 1, e, π and i. It is derived by setting $y = \pi$ in the previous equation. Since $\cos \pi = -1$ and $\sin \pi = 0$, the equation becomes $e^{i\pi} = -1$. If we bring the 1 over to the other side of the equation, we get:

$$e^{i\pi} + 1 = 0.$$

Look out for this result wherever mathematicians proclaim their geekiness!

Another consequence of all this is that, since we can write $z = x + iy$ in terms of a modulus $|z| = r$, and an argument θ as $z = r(\cos \theta + i \sin \theta)$, the modulus–argument description of a complex number is given by $z = re^{i\theta}$.

Complex functions

A complex function $f(z)$ is simply a function of a complex number, $z = x + iy$. Since $f(z)$ is complex, it has both a real and an imaginary part, often written as $u + iv$. The theory of complex functions is frankly bizarre, generating all sorts of results that are particular to the world of complex analysis. This is because being a function of z is very restrictive; the function must be written without use of the complex conjugate $z*$. Thus the real part of z is not a complex function.

This special nature becomes particularly obvious when complex functions are iterated (see page 96). During iteration, a new number is defined as the function of the previous number, and the whole process is then repeated. The sequences produced using this approach are the subject of an area of study known as *dynamical systems*. An example of the beautiful structures created from a straightforward complex function such as $c + z^2$ is given opposite. It shows an set of points that do not tend to infinity under iteration, known as a Julia set.

Complex differentiation

The derivative of a complex function is defined in the same way as that of a real function (see page 208): it measures the way a function changes as its input changes. Thus, the derivative of f at z, if it exists, is $f'(z)$ where $f(w) - f(z)$ tends to $f'(z)(w - z)$ as the complex variable w tends to z. This means that if $f(z) = z^2$ then the derivative $f'(z)$, is $2z$, which is what we would expect.

Because of the two-dimensional nature of the limit and the particular form of the complex function, satisfying this definition imposes many more restrictions on the function than might be expected. For example, if $z = x + iy$ and $f(z) = u + iv$, then f is *complex differentiable* if and only if rules called the *Cauchy–Riemann* relations hold for the partial derivatives $\frac{\partial u}{\partial x} = \frac{\partial v}{\partial y}$ and $\frac{\partial u}{\partial y} = -\frac{\partial v}{\partial x}$. This in turn implies that u and v are *harmonic functions* that satisfy $\frac{\partial^2 u}{\partial x^2} + \frac{\partial^2 v}{\partial y^2} = 0$. This is Laplace's equation, which is among the most ubiquitous equations in mathematical physics.

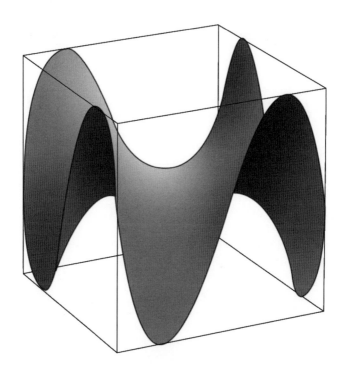

A harmonic function.

Analytic functions

Analytic functions are those complex functions that are differentiable. Since, in order to be differentiable at all, a complex function must satisfy Laplace's equation (see page 300) such functions must be differentiable not just once, but twice. We might naturally expect twice-differentiable functions to be rarer than singly differentiable ones, but in fact the opposite is the case: it is so hard for a complex function to be differentiable at all that if a function can be differentiated once, this implies it can be differentiated infinitely often. This is about as far away from the case of differentiating real functions (see page 208) as it is possible to get!

So, in the complex case, if one derivative exists then all exist. Now suppose that f and g are two analytic functions, each having a convergent Taylor series in some region of the complex plane. If the regions overlap and $f(z) = g(z)$ in the overlapping zone, then $f(z) = g(z)$ everywhere. This technique of *analytic continuation* is used in analysing the Riemann zeta function (see page 394).

Analytic continuation: the image shows overlapping regions in the complex plane. If the Taylor series of a function converges in one region and the Taylor series of another function converges in a second region, but the two functions are equal in the overlap, then they are Taylor functions of the same analytic function.

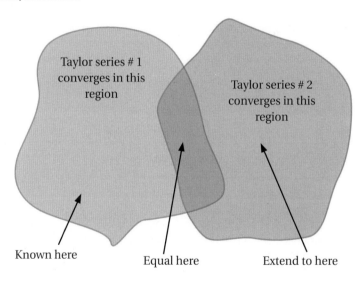

Taylor series # 1 converges in this region

Taylor series # 2 converges in this region

Known here

Equal here

Extend to here

Singularities

A singularity is any point at which a complex function becomes undefined. Complex singularities can be *removable*, if they are removed by the analytic continuation, *poles* if they behave like $\frac{1}{(z-z_0)^n}$, with $n > 0$, *essential* if the Laurent series defined below has infinitely many terms with negative powers, or *branch points* if the function is multivalued.

If near a pole $f(z)$ is defined by a power series that includes negative powers of z:

$$f(z) = \frac{a_{-n}}{(z-z_0)^n} + \ldots + \frac{a_{-1}}{(z-z_0)} + a_0 + a_1(z-z_0) + \ldots$$

This Laurent expansion is used to express complex functions that are not analytic, and cannot be represented by traditional Taylor expansions. A related representation is the Newton–Puiseux expansion, a generalized power series that can include *fractional* powers of z. This is used to create a new object, known as a Riemann surface, across which a function has a single value.

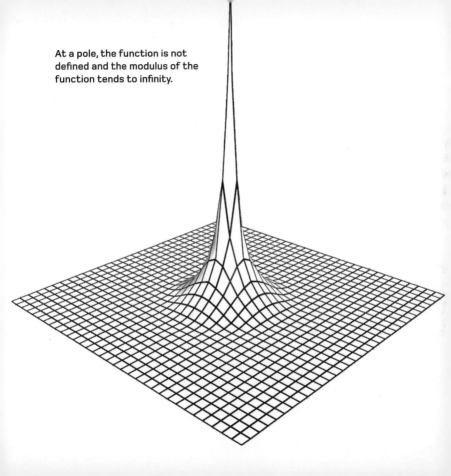

At a pole, the function is not defined and the modulus of the function tends to infinity.

Riemann surfaces

A Riemann surface is one where a multivalued function of the complex plane becomes single-valued on the surface. The natural logarithm, $\ln(z)$, of the complex number $z = |z|e^{i\theta}$, is $\ln(|z|) + i\theta$. But since $e^{2i\pi} = 1$, using Euler's identity (see page 296), $z = |z|e^{i(\theta + 2\pi)}$, so $\ln(z)$ is also $\ln(|z|) + i(\theta + 2\pi)$. In fact, $e^{2ki\pi} = 1$ for all integer values of k, so $\ln(z) = \ln(|z|) + i(\theta + 2k\pi)$ for any integer k. This is a multivalued complex function – a slightly different example is the square root of z.

The Riemann surface shown opposite removes the multivalued nature of the natural logarithm by separating the different branches of the logarithms. If we move around the central column by one rotation or 2π radians, we do not come back to the same place, as we would on a plane, and this allows the logarithm to be single-valued on the surface. The general theory of Riemann surfaces shows how to create these more complicated models of the complex plane to make different functions single-valued.

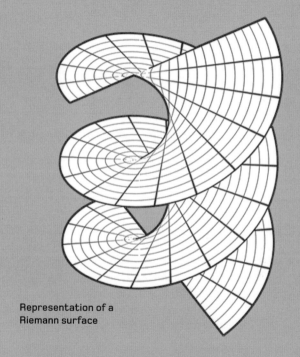

Representation of a
Riemann surface

Complex integration

As with differentiation, integration along a path in the complex plane can be defined by analogy with the two-dimensional case of a line integral (see page 236). But complex functions yield some surprising results when integrated around closed curves.

The integral of an analytic function – a complex function that is differentiable (see page 302) – around a closed curve is zero: this is Cauchy's theorem. Functions with Laurent series (see page 304) can also be integrated around a closed curve containing the pole. Here, the analytic part integrates to zero, as do all the powers of z^{-n} except z^{-1}. As a result, the only contribution comes from this term, which integrates to $\ln(z)$. The change in $\ln(z)$ around a closed curve, where the angle moves through 2π, is $2\pi i$, so this contributes $2\pi i a_{-1}$.

The coefficient a_{-1} is called a residue. So the integral of f on a closed curve equals $2\pi i$ times the sum of the residues enclosed by the curve (adding the contribution from each pole separately).

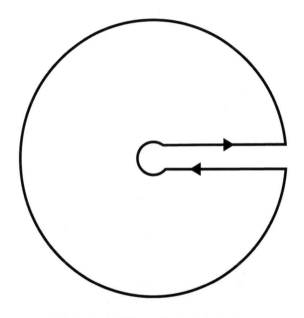

Typical curve of integration in the complex plane for the evaluation of the integral of a real function using complex methods.

The Mandelbrot set

The Mandelbrot set is a set of complex numbers which arises in the study of dynamical systems. It is the set of complex numbers C for which the origin $z_0 = 0$ does not tend to infinity under the iterated scheme $z_{n+1} = C + z_n^2$. Since $z_1 = C$ if $z_0 = 0$, another way of stating this is that iterates of the complex number C itself remain bounded. Although it is defined by the behaviour of 0 or C, a complex number being in the Mandelbrot set also gives information about its Julia set (see page 298).

Images of the Mandelbrot set are created numerically by choosing many values of C and seeing whether they grow sufficiently large under iteration to indicate that they will eventually tend to infinity, with clever tricks such as backward iteration to help fill in the detail. Those that do not are coloured black, producing the iconic and startlingly beautiful image opposite. The boundary of the Mandelbrot set is *fractal* – it has infinitely intricate, self-similar detail (see pages 338).

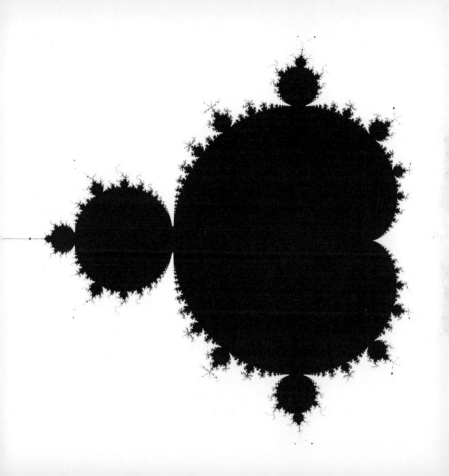

Introducing combinatorics

Combinatorics is a branch of mathematics dealing with counting. Like a poker player mentally considering the possibility of other players holding certain hands of cards, combinatorics is concerned with finding the number of objects, or possibilities of some event, without having to list all the different outcomes.

Combinatorics lies at the heart of many issues in probability, optimization and number theory. It is something of an art, and great exponents include Leonhard Euler, Carl Gauss and, more recently, the famously eccentric peripatetic Hungarian mathematician Paul Erdös.

In the past, combinatorics has been described as a discipline without a theory, reflecting a lack of unifying techniques and methods. This is beginning to change, and recent advances and successes in combinatorics suggest that it is maturing as a subject in its own right.

The pigeonhole principle

The pigeonhole principle is a simple idea with many applications. Imagine you own 101 pigeons. If you have only 100 pigeonhole boxes in which to keep them, it is obvious that at least one of your 100 pigeonholes will need to contain two or more pigeons. In more general terms, we can say that if you have n boxes and m objects with $m > n$, then at least one box will contain more than one object.

The principle can be applied to a wide range of situations. For instance, it can be used to prove that any city with more than a million non-bald inhabitants has at least two residents with exactly the same number of hairs on their heads. The proof relies on the fact that people have about 150,000 hairs, so just to be on the safe side we'll assume that the maximum number is 900,000. We therefore have a million non-bald inhabitants, the *m objects*, and 900,000 possible numbers of hairs, the *n boxes*. Since $m > n$, the pigeonhole principle tells us there must be at least two city-dwellers with the same number of hairs.

The Green–Tao theorem

The Green–Tao theorem uses methods from combinatorics to study patterns in the occurrence of prime numbers. The theorem states that arbitrarily long arithmetic progressions (see page 98) can be found using the sequence of prime numbers, though not necessarily as consecutive primes.

For example, the first three primes, 3, 5 and 7, form a sequence linked by adding the number 2 to each number. The primes 199, 409, 619, 829, 1039, 1249, 1459, 1669, 1879 and 2089 are also linked, by addition of 210. However, 2089 + 210 = 2299, which is not a prime number. So this progression breaks down after ten terms.

Such short sequences within the list of prime numbers have been recognized for years, but the theorem resisted all attempts to prove it using dynamical systems and number theory. In 2004 Ben Green and Terence Tao successfully proved the conjecture using essentially combinatorial techniques.

2 3 5 7 11 13 17 19 23 29 31 37 41 43 47 53 59 61 67 71
73 79 83 89 97 101 103 107 109 113 127 131 137 139 149 151 157 163 167 173
179 181 191 193 197 **199** 211 223 227 229 233 239 241 251 257 263 269 271 277 281
283 293 307 311 313 317 331 337 347 349 353 359 367 373 379 383 389 397 401 **409**
419 421 431 433 439 443 449 457 461 463 467 479 487 491 499 503 509 521 523 541
547 557 563 569 571 577 587 593 599 601 607 613 617 **619** 631 641 643 647 653 659
661 673 677 683 691 701 709 719 727 733 739 743 751 757 761 769 773 787 797 809
811 821 823 827 **829** 839 853 857 859 863 877 881 883 887 907 911 919 929 937 941
947 953 967 971 977 983 991 997 1009 1013 1019 1021 1031 1033 **1039** 1049 1051 1061 1063 1069
1087 1091 1093 1097 1103 1109 1117 1123 1129 1151 1153 1163 1171 1181 1187 1193 1201 1213 1217 1223
1229 1231 1237 **1249** 1259 1277 1279 1283 1289 1291 1297 1301 1303 1307 1319 1321 1327 1361 1367 1373
1381 1399 1409 1423 1427 1429 1433 1439 1447 1451 1453 **1459** 1471 1481 1483 1487 1489 1493 1499 1511
1523 1531 1543 1549 1553 1559 1567 1571 1579 1583 1597 1601 1607 1609 1613 1619 1621 1627 1637 1657
1663 1667 **1669** 1693 1697 1699 1709 1721 1723 1733 1741 1747 1753 1759 1777 1783 1787 1789 1801 1811
1823 1831 1847 1861 1867 1871 1873 1877 **1879** 1889 1901 1907 1913 1931 1933 1949 1951 1973 1979 1987
1993 1997 1999 2003 2011 2017 2027 2029 2039 2053 2063 2069 2081 2083 2087 **2089** 2099

A listing of the prime numbers up to 2100, with the sequence of primes formed by the addition of 210 highlighted.

The bridges of Königsberg

The seven bridges of Königsberg is a famous mathematical problem, whose resolution led to the development of a new discipline known as *graph theory*. In the 18th century, the Prussian town of Königsberg, now Kaliningrad, Russia, had seven bridges connecting four pieces of land across the River Pregel. The problem asked whether it was possible to tour the town crossing each bridge once and only once. Trial and error showed this was clearly very difficult, but in 1735 Leonhard Euler mathematically established that it was impossible.

By imagining each area of land as an abstract point or vertex connected by lines or edges representing the bridges, we can change the geometry of the map into a graph, removing geographical distractions. On a walk through the city, each vertex is entered and exited along an edge. To cross each bridge only once, each vertex must connect to an even number of edges. Since the vertices in fact all have odd numbers of edges, there is no path that meets our initial requirement.

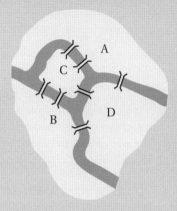

A simplified map representing the Königsburg bridge problem (above) and a graphical reprsentation (below) depicting the problem in terms of vertices and edges.

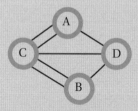

Introducing graph theory

Graph theory is the study of connections. Unlike graphs of functions, graphs in this context consist of abstract points or vertices joined by lines or edges. A sequence of vertices connected by edges is known as a *path*.

Graphs provide a useful way to analyse complex combinatoric problems. Often, solutions involve counting the number of paths of a given length within a graph, or understanding the subgraphs contained within a graph.

Many early applications of graph theory developed from the study of electric circuits, with *weightings* on edges reflecting the flow of current. Weighted graphs of flow through pipes or supply chains are also used to establish maximum flows through networks, helping to model physical or logistical processes. More recently, the internet has been viewed as a graph, and many modern models of interaction between chemicals and genes in cells are also based on graph theory.

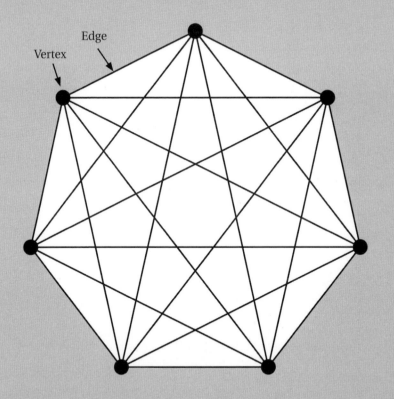

The four-colour theorem

The four-colour theorem is a classic mathematical case study. It states that the minimum number of colours that can be used to complete any map, so that no two regions or countries sharing a common border appear the same, is four.

Rephrasing the result in terms of graph theory, we can represent each region with a vertex, and connect two vertices that share a border with an edge. The problem then is to associate each vertex with a colour, so that no two adjacent vertices are the same.

As a problem requiring the analysis of many different configurations, the theorem lends itself to computer verification. In the late 1980s Kenneth Appel and Wolfgang Haken established its correctness using a computer program to check every one of 2000 or so special cases. Since then, more traditional methods have been successfully applied, and a formal analytical proof was completed in 2005.

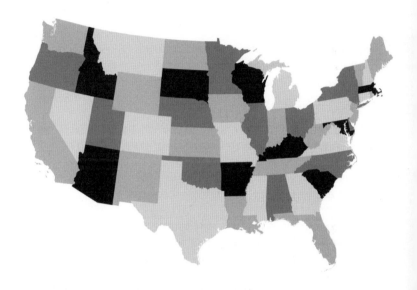

A map of the continental United States. Four different tints suffice to ensure no state shares a tint with an immediate neighbour.

Random graphs

A random graph is one in which the edges between vertices are selected by a random process. To generate a random graph, consider N vertices, and for each pair of vertices choose to create an edge, with probability p, or have no edge, with probability $1 - p$. It turns out that as N tends to infinity the properties of such graphs become independent of p and so the limiting graph is called *the* random graph.

In particular, in the random graph there will always be a path connecting any two vertices – the graph is said to be *connected*. Also, given any two finite sets of vertices there exists a vertex that is connected to all the vertices in one of the sets and none of the vertices in the other. The way that a random graph typically evolves as N increases is interesting. While N is small, the graph contains many small components and no *cycles* (non-trivial paths from a vertex back to itself). There is a cut-off for the connectivity property: if p is a bit less than $\frac{(\ln N)}{N}$ then typically there are still isolated vertices.

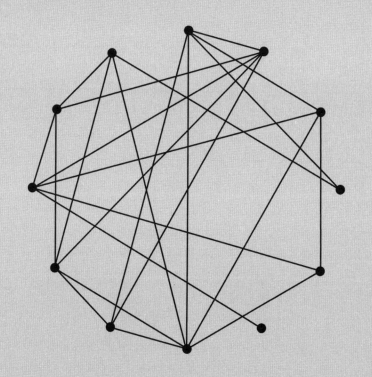

Introducing
metric spaces

Metric spaces use an abstraction of the concept of distance between objects. They are sets (see page 48) in which a distance, or metric, between elements is defined. The most familiar example is the *Euclidean metric* of three-dimensional space, in which the distance between any two points x and y is given by the length of the straight line connecting them.

More generally, a metric d and a set X are said to form a metric space if d is a real function of pairs of points in the set $d(x, y)$, that satisfies three conditions:

1. The distance between two points is non-negative, and zero if and only if the points are the same.
2. The distance between x and y is the same as the distance between y and x.
3. For any point z, the distance from x to y is less than or equal to the distance between x and z plus the distance between z and y.

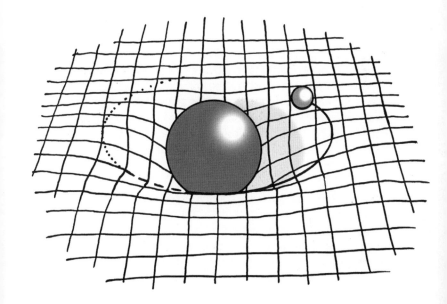

Geodesics

A geodesic is the shortest route between two points on a curved surface. On a flat surface, we know intuitively that this is a straight line. When a surface is curved, however, the shortest route may be represented by a more general curve, which minimizes a distance defined on the surface by a metric (see page 326). The most familiar non-Euclidean geodesics are great circles, such as the equator and the flight paths of long-haul aircraft.

In many cases geodesics can be determined using integration, as the minimum of a differential function describing the paths between two objects. This is how geodesics are described in Einstein's theory of general relativity, where they represent the paths of bodies through curved space–time. The fact that the shortest distances through space are actually geodesic curves can explain irregularities in the orbits of planets around the Sun, and the deflection of both light and massive objects close to black holes.

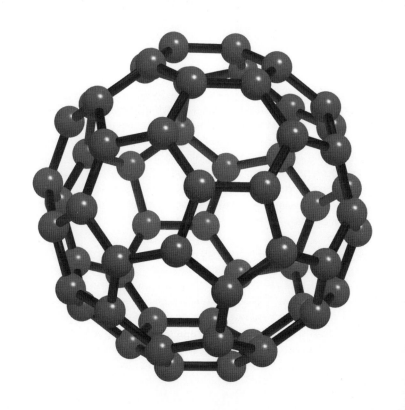

Fixed point theorems

Fixed point theorems provide conditions under which a function $f(x)$ has at least one fixed point – a point such that $f(x) = x$. Brouwer's fixed point theorem proves that, for any suitable deformation of a geometric object, the position of at least one point remains unchanged. Take two pieces of paper, scrunch one up and place it beneath the original, so that no part extends beyond. The theorem says that at least one point in the scrunched-up paper is directly below its original position.

Clearly, this only applies if we do not rip the paper in half, which, in mathematical terms, means the function f must be continuous. Similarly, the scrunched-up paper must remain inside the limits of the untouched copy, meaning that f acts on, and results in, a closed set of values. In general terms, then, if f is continuous and maps a closed set into itself, then it must have a fixed point. Similar theorems are used extensively in microeconomics, and can also be used to prove the existence and uniqueness of solutions to differential equations.

The point is directly
above where it
originally was

Manifolds

A manifold is a specific type of topological space. At a local scale, manifolds look like everyday Euclidean space, and are said to be *locally homeomorphic* to Euclidean space.

The local connection with Euclidean space provides us with a chart: coordinates with which objects in the manifold can be described. However, since this is only relevant locally, there need to be conditions to ensure that overlapping local charts are consistent with one another.

The classification of manifolds depends on the dimension of the corresponding Euclidean space (see page 108). If the manifold has five or more dimensions, then the classification is relatively straightforward, relying on a process of *surgery* whereby new structures such as holes are added to a well-understood manifold. Two- and three-dimensional manifolds have more complicated description, and four dimensions are even stranger.

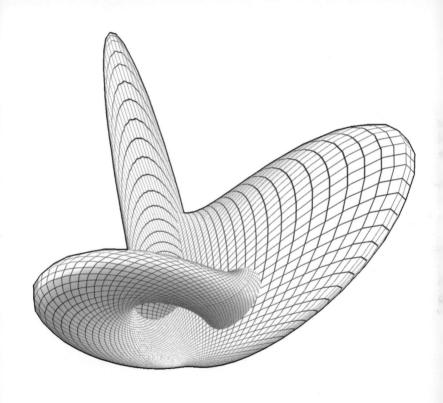

Measure theory

M easure theory provides a way of describing the size of sets in ways that generalize ideas such as length, area or volume. When you measure a set you assign a number, or weighting, to it which is an indication of its size.

Consistently defining measure via sets is hard to do, and definitions rely on the idea of a σ-algebra. This provides a way of ensuring consistency of measures, so for example the measure of a subset of a set is less than or equal to the measure of the set itself.

In many applications, statements may be true except for an exceptional set of special cases. Measure theory provides a way of quantifying the size of these exceptions. A set of measure zero is small even though it may contain uncountably many points. So if something is true except on a set of measure zero, is is said to be *true for almost all points*. For example, almost all numbers have non-terminating decimal expansions.

For a measure to be valid, it should mirror the relationship of sets. In other words the measure of the empty set should be 0, and the measure of subsets should be less than or equal to that of their parent sets.

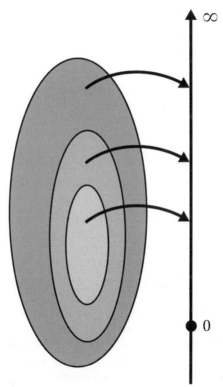

Open sets and topological spaces

Open sets are sets in which every point sufficiently close to any point in the set is also in the set. For instance, in a metric space the set of points whose distance from a given point x is strictly less than some positive number r is open and is called the open ball of radius r. Open sets are useful because they provide us with a conception of nearness of points which can be generalized to more abstract topological spaces by sidestepping the need to define a concept of distance.

Topological spaces are mathematical sets that are defined via a collection of subsets T, which are called the open sets of the space – thus the open sets are defined at the outset, not inferred from a notion of distance. The collection T of open sets must satisfy a number of specific rules:

- *T must contain both the set itself and the empty set;*
- *the intersection or overlap between any two subsets in T is also in T;*

• *the union or combination of any collection of subsets in T is in T.*

It turns out that the continuity of functions (see page 198), previously defined in terms of limits, has an equivalent definition in terms of open sets: a function f is continuous if the *preimage* of every open set is also open. The *preimage* of a set U is the set of points x whose image, $f(x)$, is in U.

Another important idea in metric spaces is *compactness*, which is an extension of the idea of a closed set. A *cover* of a space is a collection of open sets whose union comprises the entire space, and the space is *compact* if every cover has a finite subcover. That is to say, there is a finite set of open sets in the cover that also cover the original set.

This helps to define convergence. In a compact space, every bounded sequence of elements in the space has a convergent subsequence, and every compact metric space is *complete*: every Cauchy sequence (see page 88) converges to a point within the space.

Fractals

Fractals are sets which have structure on arbitrarily fine scales. Examples include the middle third Cantor set (see page 66) and the boundary of the Mandelbrot set (see page 310). The complicated shapes and surfaces of fractals are not necessarily picked up by Euclidean geometry. The middle third Cantor set is zero-dimensional as a collection of points, but it is uncountable, so has the cardinality of a line interval.

Fractals are natural objects to study from a point of view of measure theory. In particular, measure theory can be used to define an alternative 'dimension' in terms of which the middle third Cantor set has a dimension between zero and one.

The infinite intricacy of fractals is revealed if we try to cover the set with open balls of diameter r, and then let r tend to zero. If the number of balls required is $N(r)$, then as r gets smaller, then the number of balls gets bigger and for a fractal it gets much bigger as they need to cover the extra fine detail.

A series of balls of radius *r* covering the coastline of Britain. As *r* gets smaller, the number of balls required gets much larger as more detail appears—fractals require this exponential growth.

Fractal sundials

The fractal sundial is a remarkable thought experiment proposed by mathematician Kenneth Falconer in 1990. Falconer proved that it is theoretically possible to construct a three-dimensional fractal sculpture that would cast changing shadows in the shape of digits, telling the time in the style of a digital clock.

Falconer's starting point is a given sequence of thickened letters or numbers drawn in a plane, and a corresponding sequence of angles. He shows that, for every sequence of this sort, there is a fractal set such that when the angle to the sun corresponds to an angle from the given sequence, the shadow cast by the fractal onto the plane is close to the projected letters or numbers associated with that angle.

Falconer's proof is not constructive: it proves that such a sundial is possible, but does not provide a way to determine the shape of the fractal itself, and hence build a practical sundial.

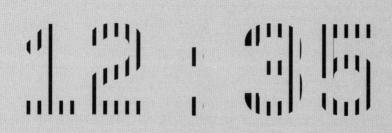

The Banach–Tarski paradox

The Banach–Tarski paradox states that a three-dimensional solid ball can be chopped up into a finite number of pieces that may then be rearranged to make two balls identical to the first. Alternatively, a small solid ball may be cut apart and reassembled into one with twice the radius. In both cases, the cut pieces are not stretched or otherwise distorted.

This clearly sounds like nonsense: cutting and moving the pieces cannot change their volume, so the volume at the beginning must equal the volume at the end. However, this is only true if the notion of volume makes sense for the pieces used in the construction. For a physical ball this is clearly the case, but for a mathematical ball, there can be other options.

The result relies on the existence of *non-measurable sets*, collections of points that do not have a traditional volume, and requires uncountably many choices in order to specify the way in which the ball is divided up.

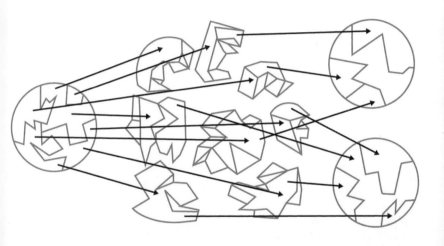

According to the Banach–Tarski paradox, it is possible to chop certain mathematical models of balls into segments that can be reassembled to make two balls identical to the original. With real balls, this is not so easy!

Introducing topology

Topology is the branch of mathematics that describes shapes and identifies when they are equivalent. The field involves considering the important properties of shapes and how they can be recognized. In topology, a doughnut and a coffee mug can be classified as 'the same', because each contains a single surface and a single hole.

Some simple examples of topological objects are the shapes that can be constructed from a sheet of paper by gluing its sides together. Glue two opposite sides together and you obtain a tube or cylinder; gluing the remaining two sides together creates a donut or torus. But two other objects, the Möbius strip (see page 346) and the Klein bottle (see page 348) can be created – in theory – by adding appropriate twists.

Topological ideas are used in computerized recognition programs and computer graphics. They can also be applied to problems such as the placement of telecoms masts.

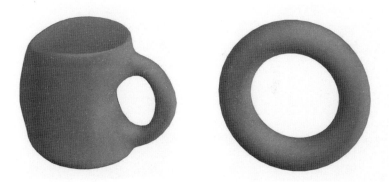

The Möbius strip

The Möbius strip is a surface with only one side and one edge. It is formed by taking a strip of paper, giving it a twist so that one side is reversed, and sticking the two ends together to form a loop.

This strip is an example of a *non-orientable surface*. *Orientability* gives a sense of whether a surface has an inside and an outside. Take a normal vector at some point, one perpendicular to the surface, and transport it continuously around the surface everywhere along a path. On a non-orientable surface such as the Möbius strip, there are paths such that when the vector comes back to the original point, it is oriented in the opposite direction to the way that it started out. Inside and outside have become confused!

Gluing two Möbius strips together along their edges gives a related object, the Klein bottle. It is not possible to do this in three-dimensional Euclidean space without tearing the paper.

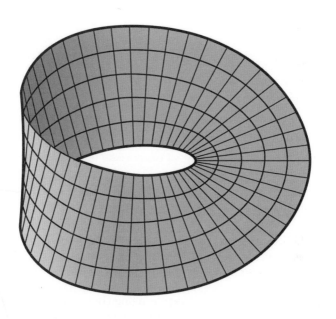

The Klein bottle

The Klein bottle is a non-orientable surface that has only one side and no edges. It is formed – mathematically – by taking a piece of paper, gluing two opposite edges together to form a cylinder, and then gluing the two remaining edges together in the opposite direction to the one that creates a torus, or doughnut.

To do this in three dimensions the surface of the Klein bottle would have to pass though itself to align the edges, but in four dimensions it can exist without self-intersections.

Unlike the Möbius strip, the Klein bottle is a *closed* surface – it is compact (see page 337) and has no boundary. Mathematicians can classify closed surfaces by counting the number of holes within the surface and determining whether it is orientable or not.

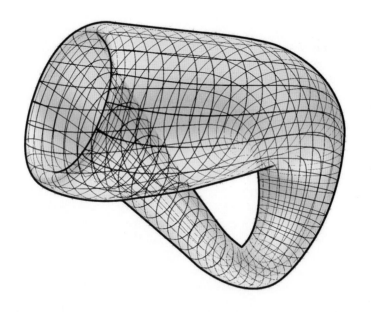

Euler characteristic

The Euler characteristic is a number that can be associated with a surface and which remains unchanged when the shape is bent or deformed. It provides a way to determine features such as the number of holes in a surface.

A polyhedron is a particularly simple closed surface consisting of a number of flat faces bounded by straight edges, which meet at vertices. Leonhard Euler noted that, for any suitably defined polyhedron, with a number of faces F, a number of edges E and a number of vertices V, $V - E + F = 2$. More general surfaces can be divided into curved faces and edges, meeting at vertices in a similar way. In the torus shown opposite, $V = 1, E = 2$ and $F = 1$, giving $V - E + F = 0$.

The number $V - E + F$ is known as the Euler characteristic of a surface. For an orientable closed surface the number of holes g, known as the *genus of the surface*, is related to the Euler characteristic through the equation $V - E + F = 2 - 2g$.

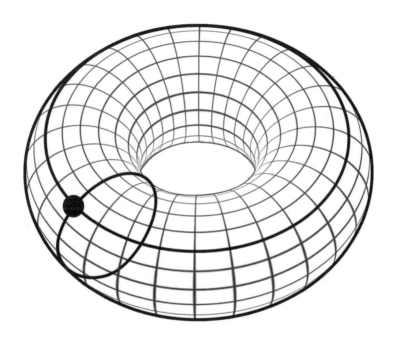

The torus: one vertex, two edges and one face.

Homotopy

Two surfaces or objects are said to be *homotopic* if one can be deformed into the other without cutting or tearing it. For instance, a coffee cup and a torus, both of which have one surface and one hole, are homotopic because each can be continuously transformed into the other.

Formally, a homotopy between two continuous functions f and g is a continuous family of transformations from one function to the other, and the spaces X and Y are said to be *homotopy equivalent* if there are continuous maps f and g, such that applying g then f is homotopic to the identity in Y and applying f then g is homotopic to the identity in X. In some sense then f and g can be seen as inverses of each other, smoothly connecting the two spaces X and Y.

Some cases, such as the horned sphere discovered in 1924 by J.W. Alexander and shown opposite, are quite surprising – this object is homotopic to the standard two-dimensional sphere!

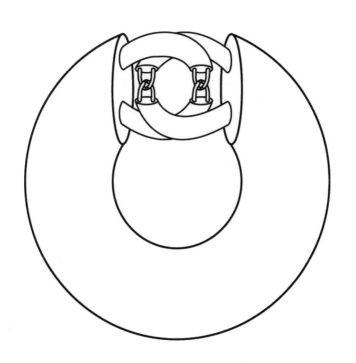

The fundamental group

A s its name suggests, the fundamental group of a topological space is a mathematical group (see page 268) associated with a topological object that characterizes the holes and boundaries of the object. It is invarient under homotopy and is based on the way that loops on the surface can be deformed.

Loops are paths within the space that start and end at the same point. Two loops are equivalent if one can be deformed into the other, and so the fundamental group encodes information about the shape of the space. It is the first and simplest of a series of homotopy groups applying to multi-dimensional spaces.

The simplest way to define a fundamental group is to fix some point x in the space X and consider all loops based at that point. Given two loops, each defining a broader class of loops in the space, we can form new classes by following one loop and then the other. In this way we create an operation on classes of loops which forms a group: the loops and this operation together form the fundamental group of the space. The fundamental group remains unchanged even if the space itself is deformed.

To give an example, consider a simple torus or doughnut ring as our space, and select a single point on the surface. From here, we can construct a loop around the perimeter of the torus enclosing the hole and a loop through the hole. These two loops are not equivalent – we cannot deform one into the other – and form templates for two classes of loop that can be manipulated to form further loops. There is a third class of loop, those that can be smoothly shrunk back to the original point, and these do not count towards the fundamental group.

The fundamental group can be used to count one-dimensional loops in the topological space, while higher-dimensional homotopy groups can be defined using spheres. In principle these provide information about the global structure of the space, but unfortunately they are very hard to calculate. Simple unchanging properties that encode information in different ways are necessary for higher dimensions (see page 254).

Betti numbers

Betti numbers are a set of numbers that describe features of a topological shape or surface, and which can be calculated using homology. Like the Euler characteristic, Betti numbers help us classify structures in terms of simple properties, such as the number of connected components, the number of holes and the number of bubbles.

Consider a piece of Swiss cheese. Important topological information would include the facts that:
- *it is a single piece of cheese, and hence is one connected component;*
- *it has n holes running through it, known as the number of topologically different non-contractible loops;*
- *there are m 'hidden' holes or bubbles inside it, the number of non-contractible three-dimensional spheres.*

These pieces of information, or their higher-dimensional equivalents, are the first three Betti numbers of the object.

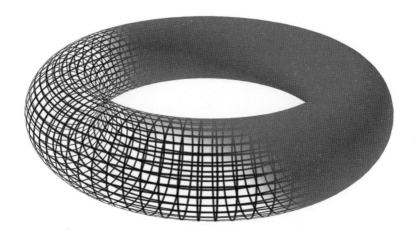

A torus like this has one connected component, two circular holes (one through the centre and one inside the surface) and one three-dimensional void (the one inside the surface). This gives its first three Betti numbers as 1, 2, 1.

Thurston's geometrization theorem

Thurston's geometrization theorem allows the classification of closed three-dimensional surfaces. In 1982, Bill Thurston listed eight known classes of three-dimensional manifolds (3-manifolds), each of which can be linked to a different definition of distance on the surface. Thurston conjectured that every other three-dimensional surface could be obtained by 'sewing together' examples of these eight basic types.

Each of Thurston's eight classes is linked to a Lie group (see page 278). The simplest is linked to Euclidean geometry and contains ten finite closed manifolds, while others include spherical and hyperbolic geometries, which have not been fully classified. The way they can fit together is reflected in the structure of the fundamental group of the 3-manifold.

In 2003, Grigori Perelman proved the conjecture using an advanced technique called a Ricci flow to determine whether various geometries were equivalent.

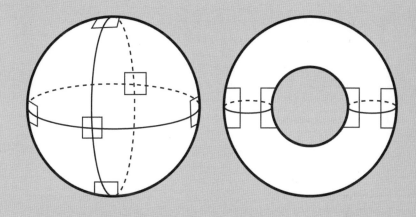

In Thurston's geometrization conjecture, three-dimensional surfaces such as spheres and doughnuts are stitched together from manifolds.

The Poincaré conjecture

The Poincaré conjecture is one of the Clay Institute
Millennium Problems (see page 404) and was the first to
be resolved — by Grigori Perelman in 2003. In simple terms, it
suggests that all three-dimensional closed manifolds with no
holes are topologically equivalent to a three-dimensional sphere.

A space has no holes (known as being simply connected) if
every loop can be contracted to a point, so the fundamental
group is trivial. In two dimensions, the only surface with this

**On any surface that
is homeomorphic
to a normal sphere,
loops can be
tightened to a point.**

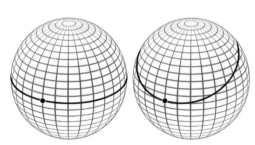

property is the surface of a topological sphere. In 1904 Henri Poincaré conjectured that this is also true in three dimensions. The issue was whether there might be some outrageous and surprising three-dimensional manifold that was simply connected, but not a sphere. Perelman proved that Thurston's geometrization theorem (see page 358) rules out this possibility, though to date he has refused to collect the million-dollar reward associated with this result.

The analogue of the Poincaré conjecture for higher dimensions was actually resolved earlier. The five-dimension problem was tackled in the 1960s by Stephen Smale, with a later improvement by Max Newman. The four-dimension situation was addressed by Michael Freedman in 1982.

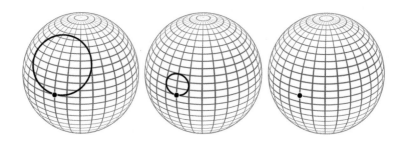

Homology

Homology is a way of measuring holes in topological spaces. It involves looking at those sets within the space that have no boundary, but are not the boundary of something else, thereby identifying the holes.

The homology groups of a space can be computed by triangulating the set: converting it into vertices, edges, triangular faces, tetrahedral volumes and so on up to higher dimensions. These can be organized to form a group structure using *boundary operators* that decompose faces into edges and so on, and a defined direction. Another approach, called cohomology, builds the higher-dimensional parts from the lower-dimensional. Depending on the problem, either approach may prove easier or offer clearer results.

Homology groups are much easier to deal with than homotopy groups. However, since there are some subtle holes that homology does not count, homotopy may still be required.

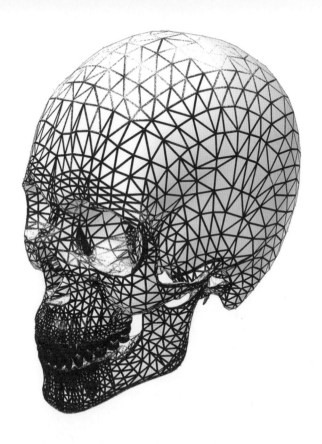

Vector bundles

Vector bundles provide a way of considering topological structures defined over a surface, rather than within it. Defining the vector bundle over a surface involves associating a vector space (see page 266) with each point on the surface. By choosing a particular element in the vector space, known as a fibre, and associating it with the point on the surface, a *vector field* is created, which can be represented by an arrow vector at each point.

Bundles provide a rich set of ways to describe manifolds. The Euler characteristic (see page 350) arises naturally in this context as a *self-intersection number* telling us about zeros of vector fields on the surface: if it is non-zero, then any continuous vector field on the manifold must have a zero somewhere. This is sometimes called the *hairy ball theorem*: the hairs correspond to a vector field on the manifold, and the existence of a zero corresponds to the fact that any way of combing the hair produces at least one crown.

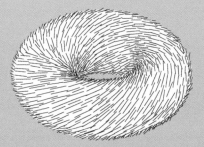

The hairy ball and hairy torus demonstrate the possible flow of vector bundles across a surface.

K-theory

K-theory was developed in the 1950s and provides a way of separating the vector bundles over a manifold into different clases – rings and groups (see pages 280 and 268). This classification leads in turn to yet another way of counting holes in a topological surface.

K-theory has parallels with cohomology, a more refined version of homology (see page 362). It has proved a very useful tool with applications in differential equations, and also provides the theoretical basis for the development of the field of non-commutative geometry – the geometry of spaces whose algebraic descriptions are non-commutative, in other words, where xy does not necessarily equal yx. In theoretical physics, K-theory plays an important role in some of the string theories that attempt to describe the fundamental particles of the Universe as vibrating multi-dimensional *strings*.

Knot theory

A knot is a closed curve embedded in three-dimensional space. Two or more such curves are known as links. Knot theory aims to describe and classify knots, considering how they should be represented, and what rules can distinguish between them.

In this context, knots are considered to be equivalent if their curves can be continuously deformed from one to another without cutting or tearing. Nevertheless, the challenge of comparing knots still has no easy solution. There are a number of knot invariants, properties that are the same for all knots of a certain type, and are unaltered by transformations, but in all known cases there are different knots that can have the same knot invariant, so they are not diagnostic.

Knot theory is useful in biology for describing configurations of DNA and related long proteins. It is also used in low-dimensional dynamical systems to determine how the periodic *orbits* of some differential equations can interact.

Logic and theorems

'Once you have eliminated the impossible, then whatever remains, however improbable, is the answer,' argued Sherlock Holmes. Holmes' method is that of the mathematician, and we use words like rigour and precision to describe the state of mind that makes deductions possible – the ability to see that all possibilities have been covered, and that there is no ambiguity and no untreated special case.

Logical *connectives*, such as *implies* or *there exists* or *for all*, are used in this book without much elaboration, but it is worth recognizing that logic is an area of mathematics in its own right.

Mathematical arguments use rules of logic, which determine how statements about properties of mathematical objects can be manipulated so that, if some elementary statements are true, then statements constructed from them are also true. But it is not just manipulation that provides meaning: the properties and objects, being abstract, require formal definition. Being precise about our deductions only makes sense if objects and their properties are accurately described.

Ideally, mathematics starts with a set of objects – primitives – and axioms – properties of these primitives. More complicated statements are then built from these using logic. Examples of such *axiomatic* systems include classical geometry (see page 108) and set theory (see page 48).

From definitions and intuition we create conjectures. These are statements that we would like to prove or disprove. A proven conjecture is called a theorem, and should be correct, accurate and precise. Theorems purport to tell us something new about the objects we are considering – something that follows logically from the definitions we started with. The Hungarian mathematician Paul Erdōs is said to have described a mathematician as a device for turning coffee into theorems.

The surprising thing about mathematics is that it appears possible to produce results that are extraordinarily non-trivial, even if they are tautological in the strict definition of that word. Although they follow logically from supposed truths, they do not become obvious without a great deal of effort.

Introducing proof

A proof is an argument that demonstrates a result, not just beyond *reasonable* doubt, but beyond all doubt. That is at least the principle. However, in practice there is neither space nor time to reduce every argument to its complete sequence of logical steps. Details may therefore be omitted as *obvious* or *trivial*, which can lead to mistakes that render the proof invalid.

It is hard to pin down precisely what constitutes a proof. For some it is a sociological construct – something that mathematicians agree performs the role of creating certainty. For others, it is a recipe that could be checked by a machine, or a Martian, that understands logical syntax.

There are several distinct strategies for formulating proofs, and different approaches can be more or less successful for any given problem. One of the arts of mathematics is finding the easiest or the most elegant path to a result.

Lewis Carroll's famous children's book *Alice in Wonderland* is littered with examples of proof and the logical fallacies that arise when its methods are not properly understood.

Direct proof

The simplest type of proof is the direct proof – one that follows a sequence of logical statements from a set of assumptions to a desired conclusion.

However, since it is almost impossible – and intolerably boring! – to write out every elementary step of a proof from the initial axioms of the field in complete detail, even a direct proof generally involves short-cuts.

The standard arguments in direct proof are sets of simple rules of inference, such as the technique known as *modus ponens*. Suppose we wish to prove a statement Q. If we can establish that *if P is true then Q is true*, that is, *P implies Q*, for some other statement P that we can already prove to be true, then this two-step proof is the same as proving Q to be true directly.

To take a simple example, suppose we wish to show that the square of every positive even number is divisible by four. Now, if a number is even and positive then it can be written as $2n$, for some positive natural number n. Its square is $4n^2$, which is divisible by four.

Here, statement P is *a positive even number can be written as 2 times a positive natural number*, and statement Q is *the square of a positive even number is divisible by four*.

This may seem trivial – it is little more than a rearrangement of the definitions of the numbers involved – but direct proof is the foundation of many proofs in mathematics.

Of course, not all methods of proof are so simple to follow, and some – proof by diagram, probabilistic proofs and mathematical induction (see page 382), for example – provoke challenging philosophical debates.

Proof by contradiction

Proof by contradiction is the mathematical version of the logical argument known as *reductio ad absurdum,* or reduction to absurdity, in which the denial of a statement leads to an absurd or nonsensical outcome. In mathematics, the absurd statement is a contradiction of something known to be true.

The argument uses the following line of reasoning:
· *To show that Q must be true, suppose that it is not true, suppose that the negation of Q is true.*
· *Use other methods of proof to demonstrate that a consequence of this assumption is a statement that is known to be false. For example, 'prove' that $0 = 1$.*
· *This shows that the initial working assumption must have been false, and that Q is therefore true.*

The proof that there are infinitely many prime numbers (see page 388) is an example of this approach.

THOECHSTE · WAS
DOEN · NIET · VALLEN

Existence proofs

Existence proofs establish that there really are objects with the properties being defined. Since mathematical objects are often abstractions, existence proofs can prevent you from expending a lot of energy investigating the properties of objects that simply don't exist, even in the abstract sense.

There are two basic classes of existence proofs. As the name suggests, a *proof by construction* produces a concrete example of the object or property, in as much as any abstract theoretical object can be called concrete. The alternative is a *non-constructive* proof – a demonstration that it is logically necessary for such an object to exist without giving a clue about examples.

Constructive proofs are fairly obvious. For example, we could ask whether there are even numbers that are divisible by 16?

The answer is yes, and the short proof is simply: 16. The longer proof is that 16 is clearly divisibly by 16 and by 2 and hence it is an even number divisible by 16. Of course, many other numbers could have been used in the proof, for instance any positive integer multiple of 16. But, for the existence proof, we only need to demonstrate one *exemplar*.

Non-constructive proofs can be quite subtle. For example, it is possible for us to show that an equation such as $9x^5 + 28x^3 + 10x + 17 = 0$ has a solution, without actually being able to say what the solution is.

Evaluating the right-hand side of the equation above with the value of x set as 0, $x = 0$, gives 17, while if $x = -1$ the result is -30. From these results we can use the intermediate value theorem (see page 202) to show that, for any value y between -30 and 17, there exists a value of x between -1 and 0 which will produce a result y from the equation. Since 0, the result we have been given on the right-hand side of the equation, lies within this range, a solution to the equation exists – a little more work shows that it is also unique – the only possible solution using real numbers.

Contrapositives and counterexamples

The negation of a statement *P*, sometimes called *not P*, is the statement which is true if *P* is false and false if *P* is true. An important rule of logic is that the statement *P implies Q* is logically equivalent to *not Q implies not P*. Sometimes it is easier to prove a link between negations than between initial statements, and this is said to be a proof by *contrapositive*.

The use of the contrapositive can only be successful if the statement to be proved is true. But in mathematical research, where inital statements might be conjectures, there is always the chance that a statement is *not* true, and there is no proof.

If this seems likely, then there are two strategies that can be taken. One is to try to logically prove the negation of *Q* instead of *Q*; the other is to find a counterexample – a single instance that contradicts the statement *Q*. For example, if *Q* is the statement *all even numbers are divisible by 4*, then 6 is a simple counterexample which disproves the statement.

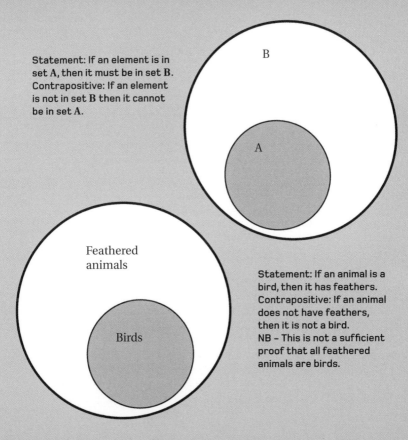

Statement: If an element is in set A, then it must be in set B.
Contrapositive: If an element is not in set B then it cannot be in set A.

B

A

Feathered animals

Birds

Statement: If an animal is a bird, then it has feathers.
Contrapositive: If an animal does not have feathers, then it is not a bird.
NB – This is not a sufficient proof that all feathered animals are birds.

Mathematical induction

Some mathematical results involve statements that depend on a natural number, so the statement to be proved is something like, *for each $n = 1, 2, 3, \ldots, P(n)$ is true*. Induction provides a way of dealing with this infinite set of statements using one theoretical idea.

Rather than establish the result for each value of n separately, mathematical induction uses the following sequence of steps:
1. *Show that the result is true if $n = 1$, i.e. prove $P(1)$.*
2. *Assume that the result is true for $n = k$, with $k \geq 1$.*
3. *Prove that if $P(k)$ is true then $P(k + 1)$ is true.*
4. *This establishes $P(n)$ for all n.*

Step 4 follows from the first three by what is called a bootstrap argument. $P(1)$ is true by step 1. Since $P(1)$ is true, then $P(2)$ is true by step 3. Since $P(2)$ is true, step 3 now proves that $P(3)$ is also true, and so on. However, philosophical problems with concepts of infinity lead some to reject inductive arguments.

INDUCTIVE METHOD TO PROVE

$$P(n): 1 + 2 + 3 + \ldots + n = \tfrac{1}{2} n(n + 1)$$

STEP 1: $P(1)$ states: $1 = \tfrac{1}{2} \times 1 \times (1 + 1)$. So $P(1)$ is true.

STEP 2: Assume $P(k)$, i.e. $1 + 2 + \ldots + k = \tfrac{1}{2} k (k + 1)$, for $k \geq 1$.

STEP 3: Show that $P(k)$ implies $P(k + 1)$:

Replace n with $(k + 1)$ in the definition of $P(n)$, to get:

$$1 + 2 + \ldots + k + (k + 1) = \tfrac{1}{2}(k + 1)(k + 2)$$

This is what we wish to prove using the assumption of step 2. Using step 2 on the sum of the first k terms shows that:

$$1 + 2 + \ldots + k + (k + 1) = \tfrac{1}{2} k (k + 1) + (k + 1)$$

But by either multiplying out the brackets or by factorizing the right-hand side as $(k + 1) \times \left(\tfrac{1}{2} k + 1 \right)$ and simplifying we obtain

$$\tfrac{1}{2} k (k + 1) + (k + 1) = \tfrac{1}{2}(k + 1)(k + 2) \text{, establishing } P(k + 1).$$

STEP 4: The general statement $P(n)$
is therefore true by induction.

Exhaustion and elimination

Proof by exhaustion breaks a problem down into subcases, and treats each separately. A historical example of such a proof is the four-colour theorem (see page 322), which was originally broken down into so many subcases that only a computer could consider them all, raising the question of whether an exhaustive computer program really constitutes a proof.

At first glance, Sherlock Holmes' process of elimination (see page 370) seems like exhaustion, but elimination actually *avoids* considering all the possibilities – it is in fact a contrapositive method (see page 380). Using exhaustive analysis of the other suspects, we prove *they* are all innocent, so we can say: *if the murderer is not Mr Ramsbottom, then none of the suspects is guilty*. The contrapositive is: *if one of the suspects is guilty, then the murderer is Mr Ramsbottom.'* The initial assumption, that we have a complete list of suspects, is often ignored, but it explains why isolated country houses feature in many detective stories.

Introducing
number theory

Number theory is the study of the properties of numbers, and often – as here – concentrates on natural numbers. Though this may seem less interesting or less important than working with real or complex numbers, the natural numbers are an intrinsic part of the way we think about the world. The sheer intellectual achievement of understanding natural numbers and their properties cannot be underestimated, and number theory involves some of the deepest questions in mathematics.

Because the natural numbers are created from the building blocks of prime numbers (see page 30), many problems in number theory concern the primes. Prime numbers are also central to the most important modern application of number theory, cryptography. The secrecy of email correspondence and bank transactions is maintained using keys based on prime factorization problems from number theory. Manipulating large primes produces codes that are easy to use but hard to crack.

```
37 — 36 — 35 — 34 — 33 — 32 — 31
 |                                |
38    17 — 16 — 15 — 14 — 13    30
 |     |                   |     |
39    18    5 — 4 — 3    12    29
 |     |    |       |     |     |
40    19    6    1 — 2    11    28
 |     |    |             |     |
41    20    7 — 8 — 9 — 10    27
 |     |                         |
42    21 — 22 — 23 — 24 — 25 — 26
 |
43 — 44 — 45 — 46 — 47 — 48 — 49 ...
```

The Ulam spiral is a remarkable pattern in prime numbers. When numbers are laid out in a simple rectangular spiral, primes show a marked tendency to lie along diagonals.

Euclid's proof of the infinite primes

The proof that there are infinitely many prime numbers is contained in Euclid's *Elements*, written over 2000 years ago. The most straightforward approach to proving the theorem uses proof by contradiction, in which denial of a statement leads to an absurd or contradictory outcome. We start, therefore, by supposing that there are precisely N primes, which can be listed $p_1, \ldots, p_N$, where N is a finite number. Now consider the number x which is the product of the N primes plus 1, that is, $x = (p_1 \times p_2 \ldots \times p_N) + 1$.

Dividing x by any of $p_1, \ldots, p_N$ will leave remainder 1, so x is not divisible by any of our finite list of primes. But since all non-prime numbers can be expressed as a product of primes (see page 30) this implies that the only divisors of x are 1 and x itself. Therefore, x must be prime. But in that case, our list of N primes was not complete. This contradicts our initial supposition, and shows that there are, in fact, infinitely many primes.

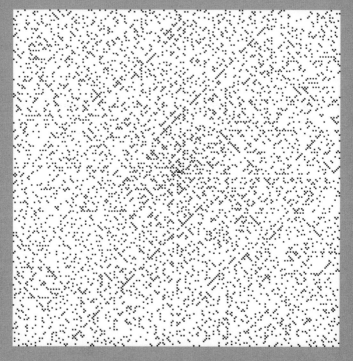

A large Ulam spiral plots the position of 40,000 numbers, with primes shown as black dots.

Twin primes

Twin primes are pairs of prime numbers that are consecutive odd numbers: that is, they are separated by two. Consider just the first few primes, for example: 2, 3, 5, 7, 11, 13, 17, 19, 23, 29, 31, 37, 41, 43, 47, 53… Here 11 and 13, 17 and 19, 29 and 31, and 41 and 43 are twin prime pairs, while 3, 5 and 7 form a prime triplet!

Numerically it has been established that there are some 808,675,888,577,436 twin prime pairs below the value of 10^{18}, and most mathematicians believe the *twin prime conjecture* that there are infinitely many twin prime pairs, although this is so far unproven.

Other pairs of primes can be constructed by analogy with the twin primes – *cousin primes* are those separated by four, while *sexy primes* are those separated by six. *Poignac's conjecture* proposes that, for any even natural number k, it is possible to find an infinite number of prime pairs separated by k.

	2	3		5		7				11		13				17		19	
		23						29		31						37			
41		43				47						53						59	
61						67				71		73						79	
		83						89								97			
101		103				107		109				113							
						127				131						137		139	
								149		151						157			
		163				167						173						179	
181										191		193				197		199	
										211									
		223				227		229				233						239	
241										251						257			
		263						269		271						277			
281		283										293							
						307				311		313				317			
										331						337			
						347		349				353						359	
						367						373						379	
		383						389								397			

Prime number theorem

The prime number theorem describes the way in which prime numbers are distributed. It states that the number of primes less than any real number x is approximately equal to $\frac{x}{\ln x}$.

Using tables of known primes, Carl Gauss was able to guess that the density of the primes is approximately $\frac{1}{\ln x}$. This means the probability of finding a prime in some small range of width d around x is roughly $\frac{d}{\ln x}$. If this is true, then the total number of primes less than x is roughly the integral of the density $\int_2^x \frac{dt}{(\ln t)}$, which is roughly of the order of $\frac{x}{\ln x}$.

The graph opposite shows that the lower line of $\frac{x}{\ln x}$ is a reasonable approximation to the upper curve of the actual number of primes less than x. But it turns out that an exact result is possible using an expression called the Riemann zeta function.

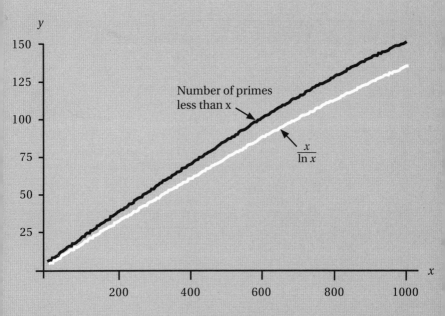

Riemann zeta function

The Riemann zeta function is intimately related to the distribution of prime numbers. It is an infinite series equal to the sum of 1 over each of the positive integers raised to a power s. This can also be expressed as a product over the primes, using a formula known to Leonhard Euler:

$$\zeta(s) = 1 + \frac{1}{2^s} + \frac{1}{3^s} + \ldots = \prod_{p \, \text{prime}} \left(1 - \frac{1}{p^s}\right)^{-1}$$

where $\prod$ indicates the multiplication of the different factors.

Using the technique of analytic continuation (see page 302), *zeta*, ζ, can be extended to an *analytic function* in which s is a complex number, $s \neq 1$. With further effort, the equation shown opposite can be established. This is striking since it is an exact relationship between the sum of the natural logarithms of primes less than x, x itself, and x^z, where the zeta function of z is zero. Hence, a knowledge of when the zeta function results in zero provides a complete description of the primes less than x.

$$\sum_{p \text{ prime}, m \geq 1, p^m \leq x} \ln p =$$

$$x - \sum_{z : \zeta(z) = 0} \frac{x^z}{z} - \frac{\zeta'(0)}{\zeta'(0)}$$

Riemann hypothesis

The Riemann hypothesis is a conjecture involving the circumstances under which the Riemann zeta function equals zero. German mathematician Bernhard Riemann initially established that there are trivial zeros for the negative even integers -2, -4, -6 and so on, which do not contribute much to the overall series. He then proposed the hypothesis that the remaining zero values all include a real part equal to $\frac{1}{2}$. This means they should lie on a line expressed as $\frac{1}{2} + ix$, where x is a real number and i is $\sqrt{-1}$. The graph opposite shows that the first non-trivial zeros arise for x-values of -14.135 and $+14.135$.

The Riemann hypothesis is one of the Clay Mathematics Institute Millennium Problems (see page 404), and also appears on David Hilbert's list of the 23 major unsolved problems in mathematics (page 68). Though the first 10 trillion zeros have been proved to appear along the $\frac{1}{2} + ix$ line, the general conjecture has yet to be proven.

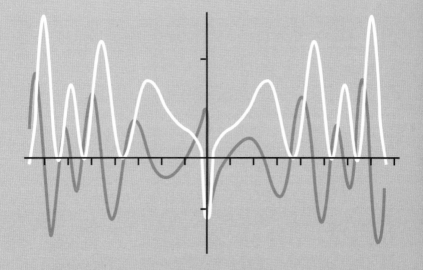

A plot of the real (white) and imaginary (grey) parts of the Riemann zeta function for $\frac{1}{2} + ix$. A Riemann zero occurs when both curves are simultaneously zero.

Pythagorean triples

Three whole numbers a, b and c form a Pythagorean triple if they satisfy the equation $a^2 + b^2 = c^2$. So (3, 4, 5) is a Pythagorean triple, since $3^2 + 4^2 = 9 + 16 = 25$, or 5^2.

It is obvious that there are infinitely many Pythagorean triples, since multiplying each component of a triple by the same factor results in a new triple. If we restrict ourselves to those triples in which the three numbers have no common divisor or factor, we can show that these are also infinite in number.

These so-called primitive Pythagorean triples can also be constructed in an elegant way. Choose positive whole numbers x and y with $x > y$, and set $a = x^2 - y^2$ and $b = 2xy$. Then $a^2 + b^2 = (x^2 - y^2)^2 + 4x^2y^2 = (x^4 - 2x^2y^2 + y^4) + 4x^2y^2 = x^4 + 2x^2y^2 + y^4 = (x^2 + y^2)^2$. The triple $(x^2 - y^2, 2xy, x^2 + y^2)$ is therefore a Pythagorean triple, and is primitive if x and y have no common factors. With a bit more work it can be shown that every primitive Pythagorean triple can be written in this form.

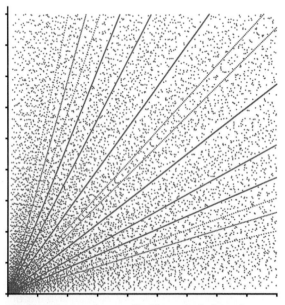

Like the prime numbers on the Ulam spiral, plots of
Pythagorean triples also reveal remarkable structure.

Fermat's last theorem

Fermat's last theorem states that no three positive integers a, b and c can satisfy the equation $a^n + b^n = c^n$ where $n \geq 3$. This is a natural extension of the Pythagorean triples, in which $n = 2$. French mathematician Pierre de Fermat jotted the theorem down as a marginal note to a mathematical textbook in 1637. Tantalizingly, he claimed that he had a method of proving it (opposite), but if that proof truly existed, it has never been discovered, though he did leave a proof for $n = 4$.

Three hundred and fifty years, and a great deal of inventive mathematics later, Andrew (now Sir Andrew) Wiles announced a proof at the Isaac Newton Institute in Cambridge. Though there was a problem with his original proof, it was soon plugged, and the final version was accepted in 1995. Wiles' approach was based on the theory of elliptic curves (see page 402), establishing that if higher triples did exist, they would contradict another major conjecture of the time. In proving *that* conjecture true, Wiles also resolved the Fermat problem.

'It is impossible to separate a cube into two cubes, or a fourth power into two fourth powers, or in general, any power higher than the second, into two like powers. I have discovered a truly marvelous proof of this, which this margin is too narrow to contain.'

Pierre de Fermat

Rational points on a curve

Rational points are numbers or values of a function that can be expressed as a ratio of two natural numbers. The identification of rational points on *elliptic curves* is important to the solution of Fermat's last theorem (see page 400).

Dividing the Fermat relation $a^n + b^n = c^n$ by c^n gives $\left(\frac{a}{c}\right)^n + \left(\frac{b}{c}\right)^n = 1$. If solutions to this equation exist, they should correspond to points on a curve $x^n + y^n = 1$, where x and y are rational numbers. For the curve $x^2 + y^2 = 1$ there are infinitely many rational points, and so the expression $a^2 + b^2 = c^2$ has infinitely many solutions, the infinite Pythagorean triples. For values of n above 2, however, things get more complicated.

This correspondence between rational points on curves and integer solutions of equations has led to a closer study of the way continuous curves intersect rational points. For simple curves there are either infinitely many rational points, or none. More complicated curves have finite numbers of points.

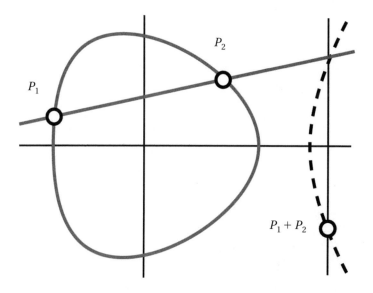

A commutative group can also be associated with an elliptic curve.
The line connecting two points intersects the curve at a third,
and the reflection of this point in the x-axis gives the group
composition of the two points.

The Birch and Swinnerton-Dyer conjecture

The Birch and Swinnerton-Dyer conjecture is an unproven statement that is one of the Clay Mathematics Institute Millennium Problems. In the same way that the Riemann zeta function counts up the number of primes, the conjecture states that there should be a similar power series that counts up the rational points of an elliptic curve.

More precisely, given an elliptic curve Bryan Birch and Peter Swinnerton-Dyer showed how to define a power series with coefficients $\frac{a_n}{n^s}$, whose behaviour at $s = 1$, they conjectured, determines whether there are infinitely many rational points or a finite number.

Although generally unproven so far, the conjecture is known to be true for several special cases. It is central to the understanding of how far functions like this can be used to determine number theoretic properties.

Clay Institute Millennium Problems

P versus NP

The Hodge conjecture

The Poincaré conjecture ✓

The Riemann hypothesis

Yang–Mills existence and mass gap

Navier–Stokes existence and smoothness

The Birch and Swinnerton-Dyer conjecture

Langlands programme

The Langlands programme is a collection of conjectures linking topics in number theory and group theory, with the potential to unify many areas of mathematics that have long been thought of as fundamentally separated. First proposed by Canadian mathematician Robert Langlands in the 1960s, the conjectures take the form of a dictionary of correspondences, suggesting that if some result is true within one theory, then an analogous result is true within the other.

The final work leading to the proof of Fermat's last theorem (see page 400) effectively resulted from following the Langlands programme. However, whilst there has been encouraging progress in this and some other directions, many other strands remain open and unproven. Nevertheless, the Langlands programme is certainly one of the great unifying themes of modern mathematics.

Glossary

Associative

An operation 'o' defined on two elements of a set is associative if $a \circ (b \circ c) = (a \circ b) \circ c$ for any three elements a, b and c of the set.

Calculus

The study of functions using limits to explore rates of change (differentiation) and sums or areas (integration).

Commutative

An operation 'o' defined on two elements of a set is commutative if $a \circ b = b \circ a$ for any two elements a and b of the set.

Complex number

A 'number' of the form $a + ib$ where a and b are real numbers and i is the square root of minus one; a is the real part and b is the imaginary part of the complex number.

Conic sections

A family of geometric curves that can be obtained by intersecting a plane with a (right circular) cone. Circles, ellipses, parabolas and hyperbolas are all conic sections.

Continuity

A function is continuous if it can be drawn without lifting pencil from paper. This means that the limit of the function, evaluated on a sequence of points tending to some point, is equal to the value of the function at that point.

Convergence

The property of tending towards a limit.

Countable

A set that can be written as a list (possibly infinite). The elements can be paired off with a subset of the natural numbers.

Derivative

The function obtained by differentiating a differentiable function.

Differentiation

The process of finding slopes or the rate of change of a function by considering the limits of the change in the function divided by the change in the variable.

Distributive

Given two operations 'o' and '×' defined on pairs of elements in a set, then × is *left distributive* over o if $a \times (b \circ c) = (a \times b) \circ (a \times c)$, and *right distributive* if $(a \circ b) \times c = (a \times c) \circ (b \times c)$ for any three elements a, b and c of the set; × is said to be distributive over o if it is both left and right distributive.

Ellipse

A closed curve that can be written in the form $x^2/a^2 + y^2/b^2 = 1$ for positive integer constants a and b. If $a = b$ the curve is a circle.

Exponential function

The function obtained by raising Euler's constant e to the power of x.

Fractal

A set with structure on all scales, so that however close you look new features emerge.

Function

A rule assigning a value (in the range or image of the function) for any value (in the domain of the function). Often denoted $f(x)$.

Group

A natural abstract algebraic structure. Given an operation 'o' defined on two elements of a set G, then G is a group if four conditions hold: $a \circ b$ is in G for every a and b in G (closure); o is associative on G; there exists e in G such that $a \circ e = a$ for all a in G (identity); and for all a in G there exists b in G such that $a \circ b = e$ (inverse).

Hyperbola

A curve that can be written in the form $x^2/a^2 - y^2/b^2 = 1$ for positive integer constants a and b.

Image

The set of all values that a function or map can take when evaluated on a given domain.

Imaginary number

A non-zero complex number with zero real part, i.e. a number of the form ib with b not equal to zero.

Integer

A whole number, including the negative numbers.

Integral

The result of integrating a function.

Integration

The process of summing areas using calculus.

Kernel

The set of vectors that map to the zero element of the vector space.

Limit

The value that a sequence tends to if it converges, so that for any desired precision, after some stage in the sequence, all subsequent terms are within that precision of the limit.

Measure

A function associated with certain subsets of a set, which can be used to determine a generalized size of different subsets. Measures are important in (advanced) integration and probability theory.

Metric

A non-negative function on points in a space that can act as a distance. If d is a metric then $d(x, y) = 0$ if and only if $x = y$, $d(x, y) = d(y, x)$ and $d(x, z)$ is less than or equal to $d(x, y) + d(y, z)$ for all x, y and z. Metrics can also be constructed by integration.

Natural number

A whole or counting number, so the set of natural numbers is $\{0, 1, 2, 3, \ldots\}$, including zero but not including infinity. Some people do not include zero in their definition, but we call the set $\{1, 2, 3, \ldots\}$ the positive integers.

Parabola

A curve that can be written in the form $y = ax^2 + ax + c$ where a, b and c are real and a is non-zero.

Prime number

A positive integer greater than 1 whose only divisors are 1 and the number itself.

Rational number

A number that can be written as an integer divided by a non-zero integer, i.e. a/b where a and b are integers with b not equal to zero.

Real number

A number that is either rational or the limit of a sequence of rationals. Every real number can be written as a decimal number.

Sequence

An ordered list of numbers.

Series

A possibly infinite sum of terms.

Set

A collection of objects, called the elements of the set. A fundamental way of grouping objects in mathematics.

Taylor series

The Taylor series of a (sufficiently nice) function about a point x_0 is a power series in terms involving $(x - x_0)^n$ with $n = 0, 1, 2, 3, \ldots$ which converges for x sufficiently close to x_0.

Uncountable

An uncountable set is a set that is not countable: that is, no list (finite or infinite) could contain all the elements of the set.

Vector

An object with direction and magnitude. A vector can be identified with a set of Cartesian coordinates $(x_1, \ldots, x_n)$ in Euclidean space or as a linear combination of basis elements in more abstract vector spaces.

Vector space

An abstract space of vectors which satisfy some rules of combination (vector addition) and scaling (multiplication by a non-vector constant).

Index

Quercus Publishing Plc
55 Baker Street
7th Floor, South Block
London
W1U 8EW

First published in 2012

A catalogue record of this book is available from the British
Library

UK and associated territories
Hardback edition: 978 0 85738 616 8
Paperback edition: 978 1 78087 369 5

Printed and bound in China

10 9 8 7 6 5 4 3 2 1